I0493711

Aging Assessment of Safety-Related Fuses Used in Low- and Medium-Voltage Applications in Nuclear Power Plants

U.S. Nuclear Regulatory Commission
Office of Nuclear Reactor Regulation
Washington, DC 20555-0001

AVAILABILITY OF REFERENCE MATERIALS
IN NRC PUBLICATIONS

NRC Reference Material

As of November 1999, you may electronically access NUREG-series publications and other NRC records at NRC's Public Electronic Reading Room at www.nrc.gov/NRC/ADAMS/index.html.
Publicly released records include, to name a few, NUREG-series publications; *Federal Register* notices; applicant, licensee, and vendor documents and correspondence; NRC correspondence and internal memoranda; bulletins and information notices; inspection and investigative reports; licensee event reports; and Commission papers and their attachments.

NRC publications in the NUREG series, NRC regulations, and *Title 10, Energy*, in the Code of *Federal Regulations* may also be purchased from one of these two sources.
1. The Superintendent of Documents
   U.S. Government Printing Office
   Mail Stop SSOP
   Washington, DC 20402–0001
   Internet: bookstore.gpo.gov
   Telephone: 202-512-1800
   Fax: 202-512-2250
2. The National Technical Information Service
   Springfield, VA 22161–0002
   www.ntis.gov
   1–800–553–6847 or, locally, 703–605–6000

A single copy of each NRC draft report for comment is available free, to the extent of supply, upon written request as follows:
Address: Office of the Chief Information Officer,
           Reproduction and Distribution
             Services Section
           U.S. Nuclear Regulatory Commission
           Washington, DC 20555-0001
E-mail: DISTRIBUTION@nrc.gov
Facsimile: 301–415–2289

Some publications in the NUREG series that are posted at NRC's Web site address www.nrc.gov/NRC/NUREGS/indexnum.html are updated periodically and may differ from the last printed version. Although references to material found on a Web site bear the date the material was accessed, the material available on the date cited may subsequently be removed from the site.

Non-NRC Reference Material

Documents available from public and special technical libraries include all open literature items, such as books, journal articles, and transactions, *Federal Register* notices, Federal and State legislation, and congressional reports. Such documents as theses, dissertations, foreign reports and translations, and non-NRC conference proceedings may be purchased from their sponsoring organization.

Copies of industry codes and standards used in a substantive manner in the NRC regulatory process are maintained at—
         The NRC Technical Library
         Two White Flint North
         11545 Rockville Pike
         Rockville, MD 20852–2738

These standards are available in the library for reference use by the public. Codes and standards are usually copyrighted and may be purchased from the originating organization or, if they are American National Standards, from—
         American National Standards Institute
         11 West 42nd Street
         New York, NY 10036–8002
         www.ansi.org
         212–642–4900

NUREG-1760

Aging Assessment of Safety-Related Fuses Used in Low- and Medium-Voltage Applications in Nuclear Power Plants

Manuscript Completed: January 2002
Date Published: May 2002

Prepared by
R. Lofaro, M. Villaram

Brookhaven National Laboratory
Energy Sciences and Technology Department
Upton, NY 11973-5000

D. Nguyen, NRC Technical Monitor

Prepared for
Division of Engineering
Office of Nuclear Reactor Regulation
U.S. Nuclear Regulatory Commission
Washington, DC 20555-0001
NRC Job Code J2831

ABSTRACT

An aging assessment of safety-related fuses used in commercial nuclear power plants has been performed to determine if aging degradation is a concern for these components in older nuclear power plants. This study is based on the review and analysis of past operating experience, as reported in the Licensee Event Report database, the Nuclear Plant Reliability Data System, and the Equipment Performance and Information Exchange System database. In addition, documents prepared by the Nuclear Regulatory Commission that identify significant issues or concerns related to fuses have been reviewed, and one fuse manufacturer was visited to obtain their insights. Based on the results of the aforementioned reviews, predominant aging characteristics are identified and potential condition monitoring techniques are discussed.

TABLE OF CONTENTS

                                                                          Page No.

ABSTRACT iii

TABLE OF CONTENTS v

LIST OF FIGURES vii

LIST OF TABLES viii

EXECUTIVE SUMMARY ix

1.0   Introduction 1

2.0   Description and Operation of Fuses 3
        2.1   Low-voltage Fuses 3
        2.2   Medium-voltage Fuses 4

3.0   Stressors, Potential Failure Mechanisms, and Failure Modes 7
        3.1   Stressors 7
        3.2   Failure Modes and Effects 7
        3.3   Potential Failure Mechanisms 8

4.0   Review of NRC Documents 13
        4.1   Part 21 Report Concerning Cracked Fuse Ferrules -
                Accession #9612170255 13
        4.2   Part 21 Report Concerning Cracking of Fuse Elements -
                Accession #9604030315 13
        4.3   Information Notice 96-23, "Fires in Emergency Diesel Generator Exciters
                During Operation Following Undetected Fuse Blowing" 14
        4.4   Information Notice 93-87, "Fuse Problems with Westinghouse 7300 Printed
                Circuit Cards" 14
        4.5   Information Notice 91-78, "Status Indication of Control Power for Circuit Breakers
                used in Safety-Related Applications" 14
        4.6   Information Notice 91-51, "Inadequate Fuse Control Programs" 15
        4.7   Information Notice No. 87-42, "Diesel Generator Fuse Contacts" 15
        4.8   Information Notice 87-62, "Mechanical Failure of Indicating-Type Fuses" 15
        4.9   Information Notice 87-24, "Operational Experience Involving Losses of
                Electrical Inverters" 16
        4.10   Information Notice 86-87, "Loss of Offsite Power Upon an Automatic Bus
                Transfer" 16
        4.11   Information Notice 85-51, "Inadvertent Loss or Improper Actuation of Safety-
                Related Equipment 16
        4.12   Information Notice 83-50, "Failures of Class 1E Safety-Related Switchgear
                Circuit Breakers to Close on Demand" 17
        4.13   NRC Weekly Information Report - Week Ending July 4, 1997 17

4.14 Evaluation of Events from NRC Documents 17

5.0 Review of Operating Experience 23
   5.1 LER Operating Experience 23
   5.2 NPRDS Operating Experience 28
   5.3 EPIX Operating Experience 36
   5.4 Trend Analysis of Operating Experience 40
      5.4.1 LER Data Trend Analysis 40
      5.4.2 NPRDS Data Trend Analysis 42
      5.4.3 EPIX Data Trend Analysis 46

6.0 Manufacturer's Insights 47

7.0 Summary and Conclusions 51
   7.1 Summary 51
   7.2 Conclusions 54

8.0 References 55

Appendix A: Fuse Types, Categories and Classification 57

LIST OF FIGURES

Page No.

Figure 1 Sample low-voltage fuse construction 4
Figure 2 Schematic of expulsion type medium-voltage fuse 5
Figure 3 Schematic of one electronic medium-voltage fuse design 6
Figure 4 Distribution of age-related fuse failures per unit from LERs 24
Figure 5 Distribution of fuse failure modes from LERs 25
Figure 6 Fuse failure mechanisms from LERs 26
Figure 7 Effect of fuse failures on plant performance from LERs 27
Figure 8 Distribution of applications for failed fuses from LERs 29
Figure 9 Fuse failure detection methods from LERs 29
Figure 10 Distribution of age-related fuse failures per unit from NPRDS 30
Figure 11 Distribution of fuse failure modes from NPRDS 31
Figure 12 Fuse failure mechanisms from NPRDS 32
Figure 13 Effect of fuse failures on plant performance from NPRDS 33
Figure 14 Effect of fuse failures on system performance from NPRDS 34
Figure 15 Distribution of applications for failed fuses reported to NPRDS 35
Figure 16 Fuse failure detection methods from NPRDS 35
Figure 17 Distribution of age-related fuse failures per unit from EPIX 37
Figure 18 Distribution of fuse failure modes from EPIX 37
Figure 19 Fuse failure mechanisms from EPIX 38
Figure 20 Distribution of fuse subcomponents failed from EPIX 39
Figure 21 Distribution of applications for failed fuses from EPIX 40
Figure 22 Fuse failure detection methods from EPIX 41
Figure 23 Number of age-related fuse failures reported per year as LERs 41
Figure 24 Number of age-related fuse failures reported per year to NPRDS 43
Figure 25 Number of age-related fuse failures per unit-year for PWRs reported to
        NPRDS 43
Figure 26 Number of age-related fuse failures grouped by major subcomponent reported to
        NPRDS 45
Figure 27 Age-related fuse failures at ANO 1 & 2 grouped by unit year - NPRDS & EPIX . . . 45
Figure 28 Number of age-related fuse failures reported per year to EPIX 46

LIST OF TABLES

Page No.

Table 1 Stressors for fuses and fuse holders 7
Table 2 Failure Modes and Effects for Electric Fuses and Fuse Holders 8
Table 3 Summary of fuse events identified from a review of NRC documents 19

EXECUTIVE SUMMARY

Fuses are included in nuclear power plant electrical systems to protect circuits that directly impact plant safety, such as containment integrity protection (i.e., to limit fault damage to a containment electrical penetration). They are also used to provide isolation protection for the Class 1E portion of the electrical system (i.e., to protect Class 1E electrical equipment from faults originating in non-Class 1E equipment). In addition, fuses are installed in motor control center cubicles, control panels, or distribution panels to protect connected loads under fault or overload conditions. From these diverse applications it can be seen that fuses are an important component in nuclear plant electrical systems, and they must function properly to ensure safe plant operation.

When a fuse is located in an environment protected from heat, moisture and caustic fumes, and the fuse is applied in a circuit with an ampacity no greater than the fuse's nominal rating, the fuse should not open unless it is subject to an overcurrent condition. However, fuses are often installed in cubicles that contain heat-generating equipment, such as transformers, resistors, and coil-energized relays. These components elevate the cubicle temperature and can cause premature aging of the fuse. Exposure to long-term, elevated temperatures might potentially affect the rated capacity and the response time (warm-up time, interrupting ability) of the fuse. Fuse contacts (ferrules or blades of the fuse) may also be subjected to environmental corrosion, particularly for plants in coastal locations, where humid, salt-air environments are common. Loose, corroded, or contaminated contacts can lead to electrical arcing across the contact surfaces and may affect their ability to maintain electrical continuity.

Based on the aforementioned issues, there is concern that the potential aging mechanisms of fuses could prevent them from performing their intended function, and could affect their ability to maintain electrical continuity throughout their service life. Therefore, this study was performed for the U.S. Nuclear Regulatory Commission (NRC) to identify those aging mechanisms that exist for fuses, and determine whether they can go undetected such that multiple safety functions could be degraded at the same time and, if called upon, could potentially fail simultaneously.

To gain insights into potential age-related failures being experienced by fuses in nuclear power plant applications, a review of past operating experience was performed. Operating experience data were obtained from the Licensee Event Report (LER) database, the Nuclear Plant Reliability Data System (NPRDS) database, and the Equipment Performance Information Exchange System (EPIX) database. In addition, NRC documents were reviewed to identify issues related to the aging degradation of fuses. The results of these reviews were used to characterize aging of fuses and draw conclusions regarding their performance as they age.

The operating experience reviewed provided a number of insights into the aging characteristics of fuses used in commercial nuclear power plant applications. Specific observations made from this review are summarized in the following:

Fuse Performance

- Considering the thousands of fuses of all sizes and types that are in service at each of the 114 nuclear power plants that operated during the period of this study, the number of age-related fuse failures reported to the NPRDS, LER, and EPIX databases was relatively low. This indicates that an age-related fuse failure is an infrequent occurrence.

- The operating experience data show that fuse failures can often go undetected until the system or component is called upon to operate. However, the designed-in redundancy of the affected instruments, components, and systems allows such failures to be tolerated with little or no effect on system or plant operation. Once they are identified, fuse failures normally are corrected easily and rapidly; NPRDS data show that more than half were corrected either on the same day or by the next day after they were discovered. In addition, the data indicate that, when an age-related fuse failure is identified, the utilities typically replace the fuses in all redundant trains for that application.

- Fuses that are operated continuously at less than approximately 60% of their rated current could potentially have an unlimited life. The worst case operating conditions would be in an application for which the fuse is repeatedly cycled from zero current to 90% or more of rated current. This would expose the fuse element to potentially severe mechanical stress due to expansion and contraction.

- The data were evaluated for trends that would indicate the degree to which aging degradation of fuses is being managed. The results show no discernible trend in the number of reported fuse failures. From 1997 to the present, an average of one fuse failure per year was reported as an LER, suggesting that age-related fuse failures are currently being controlled.

Fuse Aging Characteristics

- The predominant failure mode experienced is "FUSE OPENS SPURIOUSLY," which indicates that the fusible element opened and caused an open circuit when it should not have. The failure mode "HIGH RESISTANCE/LOSS OF CONTINUITY" is also significant and represents a failure of the fuse holder. A small number of events involve "INTERMITTENT OPERATION" of the fuse, which are typically caused by loose fuse holder clips or faulty fuse holder wiring terminations. Finally, less than 1% of the events involve a "GROUND FAULT," which is usually caused by dirt or contamination on a high voltage fuse/fuse holder assembly faulting to ground.

- The predominant failure mechanism for fuses is "FATIGUE/DEGRADATION OF ELEMENT," i.e., of the fusible element or link, which leads to unexpected failure of the fuse. This is to be distinguished from the normal opening of the fusible link when the fuse is exposed to overcurrent conditions for a prescribed time. Fatigue is typically due to the degradation of the metallic fuse element over time as a result of exposure to elevated temperature, voltage transients, or short duration overcurrent conditions. It can lead to weakening of the fuse element, or a reduction in cross section, which reduces its current carrying capacity.

- A second important failure mechanism is "WEAR/FATIGUE OF FUSE CLIPS." Fatigue of the fuse holder clips can typically occur due to high temperature, mechanical stress, and repeated insertion and removal of the fuses, such as during maintenance or surveillance testing. Other less frequently cited failure mechanisms are "CORROSION/DEGRADATION OF CONTACT SURFACES," "LOOSE, BROKEN, OR DEGRADED WIRING CONNECTIONS," and "LOOSENING/WEAR OF END CAPS."

- Most fuse failures result in either "DEGRADED TRAIN/CHANNEL" or the "LOSS OF ONE OR MORE TRAINS/CHANNEL FUNCTIONS." Less frequent effects include "DEGRADED SYSTEM OPERATION," and the "LOSS OF ONE OR MORE SYSTEM FUNCTIONS." In 16% of the fuse failures reported to NPRDS, the plants indicated that there was "NO EFFECT" at all on the system in which the problem was found.

- The most common applications in which fuses fail are control power fuses for motor operated valves and dampers or solenoid operated valves. Also significant were electronic circuit card fuses for process control circuits and systems, and closely related were instrument power supply fuses. Finally, a fairly large number of fuse failures occurred in large power supply/rectifier applications, and in battery chargers/inverters/uninterruptible power supplies.

- The greatest number of fuse failures are detected during "MAINTENANCE" and "TESTING" activities, followed by "OBSERVATION" by plant operating and maintenance personnel. This is significant because it reflects proactive efforts on the part of licensees to find these failures before they can cause more serious problems. Other fuse failures are detected by "OPERATIONAL ABNORMALITIES." This category would include failures to operate when required, off-normal performance, loss of position indication, or loss of control power indication. Finally, "AUDIOVISUAL ALARMS" accounted for the detection of the remaining fuse failures. It should be noted that several reports mentioned that imaging infrared thermography surveys were used to identify incipient fuse and fuse holder failures. Licensees are taking advantage of this powerful inspection technology to detect the tell-tale hot spots that could indicate potential fuse assembly failures.

- Fuses are essentially thermal devices. As such, external elevated temperatures can influence the performance of the fuse and change its rating. The higher the external temperature, the lower the fuse rating will be.

- Temperature cycling is a potentially significant aging stressor since the fuse element will expand and contract, and could be weakened due to work hardening. In some cases, with sand filled fuse tubes, the sand can shift during expansion/contraction of the fuse element and prevent the element from returning to its original position. This can impart mechanical stresses on the element and cause it to fail prematurely.

- Moisture intrusion is another potential aging stressor. Corrosion of the external fuse holder connections is possible, which could lead to higher resistance and heat buildup.

Also, the fuses are not hermetically sealed and moisture intrusion can occur to the inner parts of the fuse. In locations with a high moisture and sulfur content, this could lead to corrosion of the internal metallic subcomponents of the fuse and result in changes in fuse performance or premature failure. This would be a concern for fuses in circuits that are not normally operating (e.g., long-term standby circuits) since there would be no continuous heat generation to drive off moisture.

- Moisture is also a concern for fuses constructed with paper and fiber tubes. Under prolonged exposure to high humidity conditions, the paper tubes can absorb moisture causing them to experience dimensional changes. This phenomenon, if severe, could actually stretch the fuse element and cause it to break, leading to premature failure of the fuse. Once the fuse tube dries out, it may appear unchanged externally, however, it will seem that the fuse element had opened due to abnormal circuit conditions.

- Migration of zinc metal at elevated temperatures can be a concern for fuses that use zinc for the fuse element. At high temperatures, the zinc can actually migrate from one section of the fuse element to another. If the zinc migrates to a designed weak-link in the element, it can increase the cross section at that point and actually increase the rating of the fuse. Migration is not a problem with copper or silver fuse elements since these metals have been found to be dimensionally stable throughout the operating range of most fuses.

Fuse Monitoring

- Testing of the fuses is limited to resistance testing and X-rays. The resistance test is used during manufacture to determine if the fuse is acceptable. A specific resistance value is associated with an acceptable fuse in new condition. These resistance values can change with degradation of the fuse element and may be useful as an indicator of aging degradation in service. X-rays are used to verify fuse element integrity and to ensure that no excess solder is inadvertently deposited on the fuse element during manufacture, which would change the fuse performance properties. X-rays are also used in failure evaluations to provide information on the root cause of the fuse opening.

- Thermal imaging is a useful technique to locate "hot-spots," which are potential degradation sites. Typical hot-spots are loose connections and degraded contacts or connectors. If a hot-spot is detected, further investigation can be performed to determine if degradation of the fuse has occurred.

Based on the information reviewed, the following conclusions are drawn regarding aging of fuses:

- This study has found that fuses are susceptible to aging degradation that can lead to failure, however, the occurrence is infrequent. In several cases the failures have had significant impact on plant performance, including loss of redundant safety systems, challenges to engineered safety features, or reactor trips. Fuse failures in non-safety systems, such as, main feedwater, condensate, and service water, can also have an impact on plant performance. However, complete loss of equipment or system function does not normally occur due to age degradation of fuses.

- The data indicate that the incidence of fuse failures is not increasing with age presently, indicating fuse aging is being managed.

- Currently, there are several methods available for monitoring fuses, namely, high precision resistance testing, which is used during manufacture to determine acceptability of the fuse, in situ visual inspection, and thermography, which is used in the field to locate hot-spots (potential degradation sites).

- Field inspections should include examination of fuse holders since these components account for a significant number of fuse failures due to loosening of the holder clips or electrical connections. Maintenance procedures should be reviewed to minimize the removal and reinsertion of fuses to de-energize components since this can lead to degradation of the fuse holders. Fuses that must be removed and inserted frequently for maintenance and surveillance should be included in periodic maintenance and inspection programs to monitor and control the effects of these repetitive activities on the fuse and fuse holder.

- Fuses with fragile elements should be identified to assure they are handled properly during maintenance or repair activities to prevent inadvertent damage to the element.

- Fuses used in applications in which they are exposed to repeated cycles from zero load to full load should be monitored since they are susceptible to premature degradation and potential early failure due to repeated expansion and contraction of the fuse element.

- Fuses constructed with paper cartridges and used in humid environments should be monitored since they are susceptible to premature degradation and potential early failure due to moisture intrusion or swelling.

1.0   Introduction

Fuses are designed and installed in nuclear power plant electrical systems to protect circuits that directly impact plant safety, such as containment integrity protection (i.e., to limit fault damage to a containment electrical penetration). They are also used to provide isolation protection for the Class 1E portion of the electrical system (i.e., to protect Class 1E electrical equipment from faults originating in non-Class 1E equipment). In addition, fuses are installed in motor control center cubicles, control panels, or distribution panels to protect connected loads under fault or overload conditions. From these diverse applications it can be seen that fuses are an important component in nuclear plant electrical systems, and they must function properly to ensure safe plant operation.

When a fuse is located in an environment protected from heat, moisture and caustic fumes, and the fuse is applied in a circuit with an ampacity no greater than the fuse's nominal rating, the fuse should not open unless it is subject to an overcurrent condition. However, fuses are often installed in cubicles that contain heat-generating equipment, such as transformers, resistors, and coil-energized relays. These components elevate the cubicle temperature and can cause premature aging of the fuse. Exposure to long-term, elevated temperatures might potentially affect the rated capacity and the response time (warm-up time, interrupting ability) of the fuse. Fuse contacts (ferrules or blades of the fuse) may also be subjected to environmental corrosion, particularly for plants in coastal locations, where humid, salt-air environments are common. Loose, corroded, or contaminated contacts can lead to electrical arcing across the contact surfaces and may affect their ability to maintain electrical continuity.

Based on the aforementioned issues, there is concern that the potential aging mechanisms of fuses could prevent them from performing their intended function, and could affect their ability to maintain electrical continuity throughout their service life. Therefore, this study was performed for the U.S. Nuclear Regulatory Commission (NRC) to identify those aging mechanisms that exist for fuses, and determine whether they can go undetected such that multiple safety functions could be degraded at the same time and, if called upon, could potentially fail simultaneously.

2.0   Description and Operation of Fuses

There are several different designs and constructions for the various fuses and fuse holders used in nuclear power plants depending on the application and rating of the fuse. In this section, the various constructions are described for both low-voltage and medium-voltage fuses, and their operation is discussed. Additional information on fuse ratings and classification is provided in Appendix A.

2.1   Low-voltage Fuses

The basic design of a low-voltage fuse includes one or more fusible elements connected on each end to a metallic end cap, which is typically constructed of copper. The end caps can have a blade or ferrule attached to it, which fits into a fuse holder. The fuse holder is wired into the circuit to be protected, thus making the fuse an integral part of the circuit. The design and dimensions of the blade or ferrule will determine which type of fuse holder will accept that particular fuse. Rejection features are incorporated into some fuse blades to prevent the use of certain fuses in applications for which they are not designed. The fusible elements are encased in a tubular enclosure called a cartridge that can be filled with a filler material, which is usually pure quartz sand. The sand is very pure, and is vibrated and compacted to ensure no voids exist that can cause discharges. The cartridges can be constructed of paper or fiberboard, however, newer designs use melamine or plastic-impregnated glass fiber or cloth tubing to mitigate moisture intrusion and withstand the high temperatures and large forces that can occur when high fault currents are interrupted. This basic design is shown schematically in Figure 1.

Under normal operating conditions, when the circuit current is at or below the rating of the fuse, the temperature inside the fuse will stabilize below the level at which the fusible element(s) will melt. When an overcurrent condition exists in the circuit, the fusible element inside the fuse will overheat and melt. The time it takes for the element to melt and break the circuit will decrease as the current level increases in accordance with the time-current characteristic for that fuse. It will also be affected by the design of the element and the local environmental operating conditions, such as elevated temperatures. As the element melts, an arc forms across the gap between the two ends of the melted element and continues to melt away the element until the gap increases to a point where the arc is extinguished. The filler material helps to extinguish the arc and absorb the heat generated. Once the arc is extinguished, the circuit is broken and any potential further circuit damage is prevented.

The fusible element is constructed of zinc, copper, or silver alloys, and may contain one or more narrow sections. These narrow sections are weak-links that are designed to melt under prescribed conditions. The number of fusible elements and their design are dictated by the specifications of the fuse. Non-current limiting fuses may use one large element, which is suitable for heavy overloads and short circuit currents. This can be combined with a time-delay element to form a "dual-element" fuse (see Figure 1). Current-limiting fuses use multiple elements with several weak-links in each instead of one large element. This use of multiple, smaller elements helps to quickly (within one-half cycle of the event) extinguish the arcs caused by high fault currents, and limit the current passed during a fault.

3

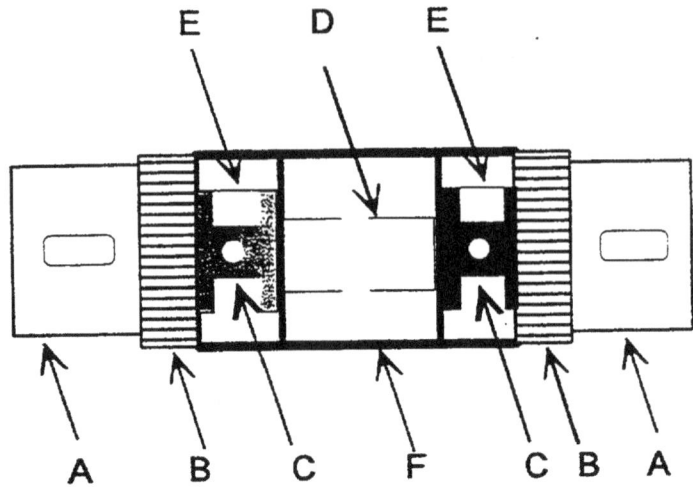

Figure 1 Sample low-voltage fuse construction

A = Fuse Blades   D = Time Delay Element
B = Metal End Caps   E = Filler Material
C = Fusible Elements   F = Fuse Cartridge

In time-delay fuses, a separate time-delay element is included in the fuse. This element can be a eutectic alloy with a low-melting temperature, as shown in Figure 1. This type of element remains firm until the temperature reaches its melting point, at which time the alloy melts suddenly opening the circuit. This time delay allows overload conditions to exist temporarily without blowing the fuse, which is desirable in certain applications, such as for electric motor starting. Other time-delay elements use a spring-loaded connecting piece that is soldered to one end of the element. At a predetermined temperature, the solder melts and the spring pulls the connector out of position, thus breaking the circuit. Another time-delay element uses a shield or button on a section of the fusible element that develops a local hot spot on the element to cause preferential melting at that point after a predetermined period of time under overload conditions.

The fuse holders into which the fuses are placed are typically constructed of blocks of rigid insulating material, such as phenolic resins. Metallic clamps are attached to the blocks to hold each end of the fuse. The clamps can be spring loaded clips that allow the fuse ferrules or blades to slip in, or they can be bolt lugs to which the fuse ends are bolted. The clamps are typically made of copper.

2.2   Medium-voltage Fuses

The design of fuses for medium-voltage applications has some similarities to low-voltage fuses, although they are usually more complex. The current-limiting medium-voltage fuse is similar in design to the low-voltage type. Multiple fusible elements containing several weak-links in each are used to rapidly extinguish the arcs caused by the interruption of high fault currents. Low-melting point solder or eutectic alloy is used in the design of some medium-voltage fusible links. Under overcurrent conditions, the solder melts creating a localized point of high resistance, leading to melting of the link at that point. Quartz sand is used as a filler material around the fusible element to help extinguish the arcs and absorb the heat generated.

Another type of medium-voltage fuse is the expulsion fuse, which is also known as the solid-material fuse or boric-acid fuse. This type of fuse generates a gas to help extinguish the arc generated by interruption of a fault current. As the gas is generated it quenches the arc and is expelled from the fuse. The design of this fuse includes an arcing rod that is attached to the fusible element. The arcing rod is spring loaded so that when the fusible element melts, it releases the rod. Spring tension then pushes the rod up through the top of the fuse cartridge and away from the broken end of the fusible element, thus increasing the arc gap. The internal surface of the fuse cartridge is lined with a solid material that generates a deionizing gas when heated by the arc. This gas helps to extinguish the arc, thus breaking the circuit. Figure 2 shows this type of fuse schematically.

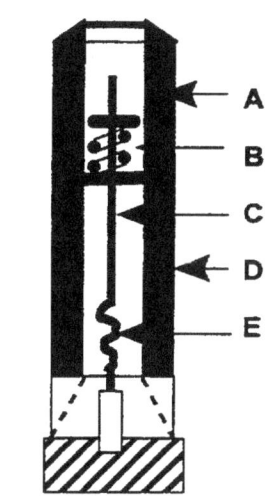

Figure 2 Schematic of expulsion type medium-voltage fuse

A = Solid gas generating material
B = Spring
C = Arcing rod
D = Fuse Cartridge
E = Fusible element

A third type of medium-voltage fuse is the electronic fuse. In this type of fuse, parallel paths are provided through which current can flow, with one of the paths containing a fusible element. The other path is the primary conducting path through which current flows during normal operation. The fusible element can be a copper ribbon, or silver element embedded in sand.

Upon an overcurrent condition, a current transformer and electronic sensing circuits sense the overcurrent and initiate a trip signal to activate a device that will disrupt the primary current path. Depending on the fuse design, the primary current path can be disrupted by activating a power cartridge, which generates a gas to drive a piston away from its contact surface, or by detonating explosive charges to force one or more selectively weakened sections of the primary conductor out of place. Once the primary path is broken, all current flows through the fusible element causing it to melt and break the circuit. One electronic fuse design is shown schematically in Figure 3.

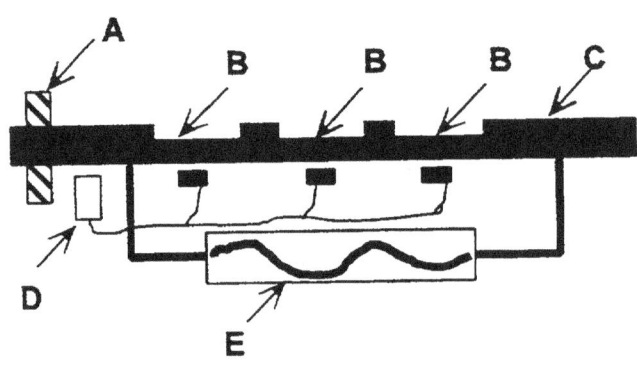

Figure 3 Schematic of one electronic
medium-voltage fuse design

A = Current sensing transformer
B = Selectively weakened breakpoints with
     pyrotechnic cutting charge below each
C = Bus bar (primary fuse conducting path)
D = Electronic sensing circuitry
E = Fusible element embedded in sand

The above provides an overview of the different fuse designs and typical construction elements. A more detailed discussion of the various types and classes of fuses is presented in Appendix A.

3.0   Stressors, Potential Failure Mechanisms, and Failure Modes

This section identifies the major stressors that contribute to the aging and degradation of fuses and fuse holders in nuclear power plants. Failure modes of electrical fuses are presented together with a list of the common effects resulting from each failure mode. The aging mechanisms that can lead to the most common failures in fuse and fuse holder subcomponents are discussed, along with some of the causes contributing to these failures, based upon an ongoing review of fuse-related failure events and technical literature.

3.1   Stressors

The stressors that affect fuses and fuse holders in nuclear power plants are shown in Table 1.

Table 1 Stressors for fuses and fuse holders

- Elevated Temperature
- Thermal Cycling
- Chemical Contamination
- Vibration
- Electrical Cycling (Power Quality)

- Moisture/Humidity
- Radiation
- Mechanical Stress
- Seismic Events
- Steam Exposure

These stressors may act independently and/or synergistically to cause degradation and eventual failure in the various parts of a fuse and fuse holder assembly, and their electrical connections. Of all the stressors listed, elevated temperature is the most important since the fuse is basically a thermally operated device. Fuses in nuclear power plant applications are normally located inside enclosed, energized electrical distribution panels, motor control centers, metal-clad switchgear, electronics racks or cabinets, instrumentation and control panels, power supplies, inverters and battery chargers, or power circuit breakers. Consequently, the fuse is always operating in a high temperature environment. In addition, ohmic or Joule heating (I^2R) is produced due to the passage of operating current through the conductive pathways of the fuse and fuse holder. Thus, heat is added not only as a byproduct of the environmental conditions, but also as a consequence of ohmic heating from the flow of electric current. It may also be seen in the subsequent discussions that many of the other stressors listed can contribute to mechanisms that cause further ohmic or Joule heating (I^2R) and heat-related degradation.

3.2   Failure Modes and Effects

On a functional level, the fuse is among the simplest of all electrical devices. When circuit conditions are within normal operating limits, the fuse serves as a part of the electrical conducting circuit and is essentially invisible to the system as a whole. Upon a sustained overload or overcurrent condition, or under the high current conditions occurring during a short circuit electrical fault, the fuse element will melt as designed, thus opening the circuit. This normal operation of the fuse will interrupt the flow of potentially damaging current into electric cables, equipment, and components.

In reviewing fuse-related operating events, depending upon the viewpoint of the evaluator, a 'blown fuse" will frequently be reported as a failure since it caused an interruption in the availability of a circuit, equipment, component, or system. Strictly speaking, however, this is not a fuse failure if the device performed as designed to protect the circuit in which it was installed. Further investigation, in most cases, will reveal that the cause of a fuse opening is typically a short circuit condition, damaged or overloaded electrical equipment, electrical transients, or other problems.

An electrical fuse is considered failed when it does not perform its protective function in accordance with its design characteristics. There are several ways in which a fuse can fail. These include spurious opening under normal rated current and voltage conditions, intermittent operation (due to a loose or faulty electrical contact or connection within the device), high internal resistance (which can accelerate heat-related degradation mechanisms), opening earlier or later than prescribed in its time current characteristic curve, or not opening at all when required. The failure modes of a fuse/fuse holder assembly are summarized in Table 2

Table 2 Failure Modes and Effects for Electric Fuses and Fuse Holders

Failure Modes	Failure Effects
• OPENS SPURIOUSLY - under normal rated current and voltage or no load at all	• Loss of function of protected circuit, component, equipment, or system
• OPENS INTERMITTENTLY - loose/faulty electrical contact	• Intermittent loss of function or trip of protected circuit, component, equipment, or system
• HIGH RESISTANCE - corrosion, oxidation, contaminated electrical contact	• Increased ohmic heating within fuse/fuse holder assembly • Accelerated heat-related degradation mechanisms in fuse assembly
• OPENS EARLY - under overcurrent conditions	• Loss of function of protected circuit, component, equipment, or system • Loss of electrical protective device selectivity
• OPENS LATE - when required	• Loss of function or damage to protected circuit, component, equipment, or system
• FAILS TO OPEN - when required	• Loss of function or damage to protected circuit, component, equipment, or system • Loss of electrical protective device selectivity

3.3   Potential Failure Mechanisms

The various subcomponents of electric fuses and fuse holders are subject to failure mechanisms that can lead to the failure of a fuse to perform its required function. The important mechanisms are discussed in the following paragraphs for the major subcomponents of the fuse assembly.

As described in Section 2, the fusible element is typically constructed of zinc, copper, or silver alloy in one or more parallel links of a specific cross section to produce the required time-current characteristics for the fuse. In a dual element current limiting fuse, a second, often spring-loaded, element is included in series with the overcurrent element to provide short circuit protection characteristics. The element is connected to the inside of the end caps of the fuse by a solder joint in most low-voltage fuses, or a brazed joint in the higher current rated low- and medium-voltage fuses. The time-current characteristic of the fuse element may be altered by several aging mechanisms acting over time to change the physical properties of the metal alloy, change the cross sectional area of the element, or weaken the tension of the spring. As discussed in Section 3.1, high temperatures in the operating environment can combine with ohmic heating to produce a higher than design thermal environment. Thermal cycling, due to operational or electrical power quality transients, or sustained operation at higher than design temperatures can cause melting of solder joints or partial melting of the element [1]. Diffusion can also lead to changes in composition and electrical resistivity across the element [2]. This can cause the element to melt faster than indicated by its time-current characteristic curve. Changes in the physical properties of the metal alloy or the cross sectional area of the element brought about by sustained high temperature or thermal cycling can also cause premature opening of the fuse.

Migration of zinc metal at elevated temperatures can be a concern for fuses that use zinc for the fuse element. At high temperatures, the zinc can actually migrate from one section of the fuse element to another. If the zinc migrates to a designed weak-link in the element, it can increase the cross section at that point and actually increase the rating of the fuse. Migration is not a problem with copper or silver fuse elements since these metals have been found to be dimensionally stable throughout the operating range of most fuses.

The soldered or brazed connection between the element and the end cap can be affected by heat, vibration, seismic events, and mechanical stresses. It is common practice to remove and tag out control power fuses as part of the tag out procedure to remove a circuit breaker, motor, or other electrical equipment from service for maintenance, testing, or surveillance. Repetitive removal, insertion, and handling of fuses over time can produce mechanical stresses and vibration that can weaken or break an aged solder connection or an aged fuse element.

Temperature cycling is a potentially significant aging stressor since the fuse element will expand and contract, and could be weakened due to work hardening. Fuses that are operated continuously at less than approximately 60% of their rated current could potentially have an unlimited life. The worst case operating conditions would be in an application for which the fuse is repeatedly cycled from zero current to 90% or more of rated current. This would expose the fuse element to potentially severe mechanical stress due to expansion and contraction. In some cases, with sand filled fuse tubes, the sand can shift during expansion/contraction of the fuse element and prevent the element from returning to its original position. This can impart mechanical stresses on the element and cause it to fail prematurely.

Moisture can enter the fuse where the fuse tube is joined to the end caps, or it can be absorbed through fuse tubes constructed of certain materials that are not impervious to water. A galvanic reaction can be established between the dissimilar metallic zinc fuse element and the copper end cap. Some control power fuses use Calcium Sulfate Dihydrate ($CaSO_4 \cdot 2H_2O$; gypsum) as a filler material around the zinc element. At temperatures above 100ºC, the hydrated water, containing sulphate and calcium ions, can be driven out of the filler material into the interior of the fuse. The presence of these electrolytes may further help to drive the galvanic reaction, especially at higher temperatures, with the zinc acting as a sacrificial anode, corroding away until the circuit opens, most likely at the solder joint to the end cap [1].

Moisture is also a concern for fuses constructed with paper and fiber tubes. Under prolonged exposure to high humidity conditions, the paper tubes can absorb moisture causing them to experience dimensional changes. This phenomenon, if severe, could actually stretch the fuse element and cause it to break, leading to premature failure of the fuse. Once the fuse tube dries out, it may appear unchanged externally, however, it will seem that the fuse element had opened due to abnormal circuit conditions.

The end caps on Class H renewable element fuses unscrew to allow replacement of the element. Vibration, thermal and electrical cycling, and mechanical stress from repetitive removal, insertion, and handling of fuses over time can cause loosening of the screw-on end caps. This can cause intermittent operation or an open circuit due to loss of electrical contact. The loose connection is also susceptible to oxidation, corrosion, and contamination by dirt, grease, or other substances resulting in a degraded electrical connection; this would produce an intermittent electrical connection or a high resistance connection with excessive ohmic (I^2R) heating that can produce or exacerbate the heat-related mechanisms described above.

The end terminals of a fuse, along with the electrical and mechanical interface between the fuse and the fuse holder, are another area that can be affected by degradation mechanisms in nuclear plant applications. Typical end terminations for low-voltage and medium-voltage fuses include ferrule-type, blade-type, and bolted-type. Miniature electronic fuses may use ferrule end caps, spade-type connectors, or axial leads that allow direct solder connection for printed circuit board surface-mount fuses.

For the ferule-type and blade-type end terminations, deterioration of the electrical contact surface is a major failure mechanism. Corrosion, chemical contamination, and oxidation of the connecting surfaces can increase the electrical resistance of the conductive path. As the condition deteriorates, the increased ohmic heating can lead to heat-related problems within the fuse, as discussed above. The heat generated can also accelerate the deterioration process. In one failure event reported to a national database [3], the high resistance heating at the end terminations became so great that the fuse burned clear at both ends, opening the circuit, and destroying the fuse and fuse holder. Loose connections at the fuse holder clips are susceptible to oxidation, corrosion, and contamination by dirt, grease, or other substances that can produce high electrical resistance, or intermittent or loss of electrical contact.

Bolted-type connections, generally provide a more secure connection than the spring pressure terminations discussed above. Nevertheless, bolted fuse connections can loosen over time due to thermal cycling, vibration, mechanical stress, or fatigue of the fasteners. These terminations can be affected by corrosion and oxidation, especially in the presence of heat, moisture, or corrosive atmospheres. Again, these processes will cause increased electrical resistance and the attendant increase in ohmic heating at the connection. Dirt, grease, and other contaminants may sometimes degrade a bolted connection and could cause corona losses at the higher voltages.

Another subcomponent of the fuse assembly that demonstrates a large number of degradation-related failures is the fuse holder. In cartridge type fuse holders for low- and medium-voltage

10

fuses, the fuse clips or fingers may be annealed and lose their spring force after just one overheating due to a high resistance connection. Annealing of a copper fuse clip may occur at temperatures as low as 93ºC, depending on the degree of cold work [1]. The cuprous oxide which forms on both copper and brass end clips, as well as copper and brass end caps and blades, contributes to a poor electrical connection. A high resistance connection between the fuse and fuse holder clips will result in arcing or ohmic (I^2R) heating at the connection that can raise its temperature to 65ºC and may triple the corrosion rate of zinc [1].

Other mechanisms that may affect the fuse-to-fuse holder connection are vibration, thermal cycling, seismic events, electrical transients, and mechanical stress. As mentioned earlier in this section, the common practice of removal and tag out of control power fuses as part of the tag out procedure to take electrical equipment out of service involves repetitive removal, insertion, and handling of fuses. This recurring procedure can loosen the spring tension on fuse holder clips and fingers, bend or mis-align fuse holder clips and fingers, fatigue rivets that hold the clips to the molded base, or crack the molded base of the fuse holder. The discussion above on the failure mechanisms of ferrule-type and blade-type end terminations covers the degradation mechanisms affecting fuse holder clips and fingers due to corrosion, oxidation, dirt and other contaminants.

Several failures were reported involving bayonet-type fuse holders with a plastic or phenolic screw on knob used for panel mounted fuses in electronic equipment or systems [3]. Loose connections at the screw-on knob can result from wear, mechanical stress, vibration, fatigue, and thermal cycling. The low-voltage glass ferrule fuses used in this type of fuse holder have zinc elements, which are susceptible to the same heat-related degradation mechanisms discussed above for the low-voltage cartridge fuses. The bayonet fuse holder's leaf spring is typically made of cold-worked brass, and the coil spring is cold-worked austenitic stainless steel. Even a single overheating of the coil spring, which provides compression on the fuse within the holder, and the leaf spring, which provides the electrical connection contact, can cause annealing and decrease their spring force [1]. The fuse then may no longer be held properly within the bayonet fuse holder; therefore, it is good practice to replace the fuse holder knob when a blown fuse has to be replaced.

Another subcomponent of fuse holders that is subject to degradation mechanisms is the wiring connection between the electrical circuit conductor and the fuse holder. This connection may include one of the following types: lug and screw, bolted compression cube, spade, or solder joint. All but the last depend on a crimping or pressure connection of the electrical conductor to establish electrical contact. All of these connections are affected by one or more of the following: thermal cycling, vibration, mechanical stress, fatigue (of fasteners and threaded connections), overheating due to high electrical resistance or electrical transients, corrosion and oxidation, contamination, and compression deformation (of the electrical circuit connectors). Solder connections eliminate many of the problems that affect the mechanical connectors, however, these joints are susceptible to similar potential failure mechanisms as those discussed above for the fuse's internal solder connections. A number of the fuse holder connector failures reported to a national database involved broken, loose, cracked, or defective solder connections between the circuit conductor and the fuse holder [3].

4.0   Review of NRC Documents

To characterize age-related failures being experienced by fuses in nuclear power plant applications, a review of past operating experience was performed. Operating experience data were obtained from the Licensee Event Report (LER) database, the Nuclear Plant Reliability Data System (NPRDS) database, and the Equipment Performance Information Exchange System (EPIXS) database. In addition, NRC documents, such as Information Notices, Bulletins, and Generic Letters were searched and reviewed to identify issues regarding safety-related fuses. The results of the NRC document review are presented in this section. Results of the data review are presented in Section 5.

A search was performed of NRC documents available through the NRC web site. The key word "FUSE" was used as the search criteria and a total of 377 documents were identified. In addition, year-by-year searches of Part 21 reports, NRC Bulletins, Generic Letters, Information Notices, and Regulatory Issue Summaries were also performed to ensure that all significant fuse-related events were identified. Thirteen events were found that are generic fuse-related events and provide insights into problems experienced with fuses in nuclear power plant applications. A number of events were also identified in the Daily Events Reports, however, these typically are isolated events that are specific to a particular plant and result in an LER. These events are captured and included in the LER search results, therefore, they are not discussed herein. Each of the generic type fuse events found from the NRC document review is summarized below.

4.1   Part 21 Report Concerning Cracked Fuse Ferrules - Accession #9612170255

In September 1996 at the Millstone Power Station, it was found that fifteen different fuse types from three different manufacturers had longitudinal cracks in the ferrule. All of the cracked fuses used brass ferrules. The manufacturers were Gould-Shawmut, Bussman, and CEFCO. It was determined that the cracks occurred as a result of stress corrosion cracking as the brass ferrule relieved internal stress introduced during the manufacturing process. Subsequent functional testing of the fuses found that the fuses met their intended function, however, ten out of forty-six fuses had the ferrules physically blow off during testing. The cracked ferrules could result in a loss of safety-related equipment function due to the ferrule coming off the fuse, which could then short out or damage other safety-related equipment. This is a potentially significant safety hazard since it could prevent safety-related equipment from performing its function. Newer fuse designs use copper or bronze as the ferrule material to mitigate this problem. The licensee determined that older fuses still in use with the brass ferrules should be replaced.

4.2   Part 21 Report Concerning Cracking of Fuse Elements - Accession #9604030315
(Note - Accession #9505150034 deals with the same issue)

In March 1995 Calvert Cliffs reported five failures of Gould-Shawmut 10-15 amp fuses over the previous several months. It was determined that these fuses had developed cracks in the fuse element. As the cracks propagated, the fuse element failed to carry current, creating the appearance that the fuse was blown. The probable cause of the fuse failures was corrosive flux residues used in the soldering process during manufacturing, which caused the fuse element to weaken, embrittle, and eventually crack. This is a potential significant safety hazard since it could prevent safety-related equipment from performing its function. Gould-Shawmut fuses manufactured after 1993 now use a different soldering process and a non-corrosive flux to avoid this problem. The licensee determined that older fuses still in use should be replaced.

4.3 Information Notice 96-23, "Fires in Emergency Diesel Generator Exciters During Operation Following Undetected Fuse Blowing"

This information notice addresses a problem experienced when blown fuses in a generator exciter were not detected. During a refueling outage at the Wolf Creek Generating Station, in September 1994, the 'A' train emergency diesel generator was undergoing post maintenance testing and balancing. After approximately one hour of sustained operation, a fire occurred in the main power potential transformer of the static exciter-voltage regulator (exciter). After the fire in the EDG-A exciter, the licensee performed electrical checks on the EDG-B exciter and found no problems. On October 11, 1994, again after about an hour of above full power operation of EDG-B, its exciter potential transformer also caught fire. After each fire, the licensee found that one of the 100-ampere fuses in the secondary circuits of the respective exciter potential transformer had blown. It was later determined that the fuses had not blown as a result of the fires, but that the blown fuses were a contributing cause of the fires. The blown fuses were not detected because these fuses had no "blown fuse" indication. Corrective actions involved the installation of blown fuse indicators on the diesel exciter cabinets. This is not an age-related event since the fuse failures were caused by improper shutdown of the generator. However, it provides insights into the types of problems encountered with fuse failures and how they are detected.

4.4 Information Notice 93-87, "Fuse Problems with Westinghouse 7300 Printed Circuit Cards"

This notice documents a problem experienced in the 1992-1993 time frame with failures of Westinghouse 7300 printed circuit cards related to the use of incorrect fuses. It was found that incorrectly sized fuses were being installed in the cards, which resulted in failure of other components on the card. In some cases, a 1.0 amp fuse was installed instead of a 0.5 amp fuse. In more severe cases, 5-amp fuses were found installed instead of 0.5 amp fuses. This notice does not represent an age-related fuse issue, however, it is included herein to provide insights into fuse problems found in past operating experience reports.

4.5 Information Notice 91-78, "Status Indication of Control Power for Circuit Breakers used in Safety-Related Applications"

This notice addresses a problem experienced with the failure of a containment spray pump to start during surveillance testing. The cause was determined to be a failure of the fuses in the closing coil circuit for the pump. The fuse holder fingers, which connect the fuses to the circuit, became deformed such that poor or no contact was made in the closing coil circuit. The circuit wiring design was such that no indication or alarm was provided for the loss of control power. This could impact plant safety since safety-related equipment could become inoperable and operators would not be aware of it. This event also demonstrates a potentially safety-significant

14

failure mechanism for fuses, namely, degradation of the fuse holder clips.

4.6   Information Notice 91-51, "Inadequate Fuse Control Programs"

This notice addresses potential problems caused by inadequate programs to control activities related to fuses. Numerous deficiencies involving fuse control programs were identified during NRC inspections, including the following:
- Inadequate root cause evaluation of blown fuses,
- Inadequate verification of design information for installed and replacement fuses,
- Inadequate identification and labeling of fuses,
- Improper coordination of fuses and circuit breakers, and
- Personnel errors

These deficiencies have resulted in significant plant safety events, including inadvertent operation or loss of vital plant equipment, loss of offsite ac power, and spurious actuation of engineered safety features. It is concluded that a well designed fuse control program along with trained personnel can reduce fuse problems significantly. This notice emphasizes the safety significance of fuses and confirms that fuse failures do have a significant impact on plant safety.

4.7   Information Notice No. 87-42, "Diesel Generator Fuse Contacts"

In April 1987 at the Browns Ferry Nuclear Plant, an explosion occurred in the electrical control cabinet during routine surveillance testing of an emergency diesel generator. The explosion resulted from a phase-to-phase short in a cable bundle in the potential transformer fuse compartment. It was determined that the insulation on the cables routing power from the potential transformer fuses to the potential transformer failed due to an over-temperature condition. The over-temperature condition is believed to be due to poor contact between the spring fingers of the transformer fuses. This is the result of misalignment or contact degradation from corrosion, pitting, or burning, which caused arcing and eventual failure. Inspection of other fuses in similar applications found three other instances of poor contact. Subsequently, the design of this fuse arrangement was modified to use a knife switch contacts.

4.8   Information Notice 87-62, "Mechanical Failure of Indicating-Type Fuses"

In the 1986-1987 time frame, four different plants reported failures of indicating type fuses. These fuses have an internal, spring-loaded indicating pin that protrudes from the end of the fuse when the fuse element separates. The fuse elements are designed to melt when the current exceeds the design load, however, in the reported cases, the fuses failed as a result of either a cold solder joint, creep, or fatigue of the element induced by the internal spring tension. The manufacturers of the fuses are Bussmann and Littelfuse. The safety-significance of the reported failures depends on the fuse application, however, several significant plant effects have resulted. Examples include the following:

- At the McGuire Station, a reactor trip occurred due to a steam generator low-low level signal when a fuse failure caused a main feedwater containment isolation valve to close,

- At the Catawba station, an auxiliary feedwater train failed to start during testing due to mechanical failure of a fuse, and

- At the Sequoyah Station, the emergency diesel generator started due to failure of a fuse in the diesel start logic circuitry

All of the above cases involved mechanical failure of indicating type fuses. The corrective actions taken were to replace the failed fuses. It was concluded that, in the event of an indicating fuse failure, additional investigation, including internal examination of the fuse, may be warranted if an electrical fault cannot be found.

4.9   Information Notice 87-24, "Operational Experience Involving Losses of Electrical Inverters"

This notice identifies potential problems with inverter failures leading to unplanned plant transients and/or inoperability or improper functioning of safety-related equipment. One of the subcomponents identified that can cause inverter failure is fuses. A stressor associated with fuse failure, and thus inverter failure, is high ambient temperature and/or humidity within the inverter enclosures, which results in accelerated aging of the fuse. Another stressor identified is voltage transients due to energizing and de-energizing various loads serviced by the inverter. These failures are considered safety significant since they can result in challenges to safety systems in the plant. The notice suggests that licensees consider monitoring temperature and/or humidity internal to the inverter enclosures and evaluate input and output voltages of the inverter unit during steady-state and transient conditions to assure that manufacturer's recommendations are being considered.

4.10   Information Notice 86-87, "Loss of Offsite Power Upon an Automatic Bus Transfer"

In January 1986 at the H.B. Robinson Station, a reactor trip occurred followed by a loss of offsite power when the emergency diesel generator output breaker was removed to install a solid-state overcurrent trip device. The breaker had just been racked out when the emergency bus tripped due to a blown potential transformer fuse. The reactor trip was caused by a fuse failure, which was attributed to a loose fuse holder. The loose fuse holder caused the fuse to blow when the breaker was racked out. This in turn caused the emergency bus to trip, followed by the loss of an instrument channel, turbine rollback, and finally a reactor trip on high pressurizer pressure. The fuse holder was subsequently replaced. This event is safety significant since it resulted in a loss of redundancy in emergency power when one of the diesels became unavailable for service. The loss of offsite power was determined to be a separate event, not related to the fuse failure.

4.11   Information Notice 85-51, "Inadvertent Loss or Improper Actuation of Safety-Related Equipment

This information notice addresses potentially significant reactor safety problems that have been caused by the normal practice of removing fuses or of opening circuit breakers for personnel protection during maintenance and plant modification activities. A potential problem can occur

16

when the effects of electrical power interruption on all circuits powered by the fuse or breaker are not fully reviewed in advance. Errors in the review have resulted in unknowingly disabling safety systems and also have caused inadvertent actuation of safety systems. Corrective actions recommended are identification of effects on plant equipment or systems, independent verification of the evaluation of effects, and utilization of the nearest local fuse or circuit breaker to minimize the number of systems affected. This notice underscores the safety importance of fuses and the potential problems that can arise due to fuse failures. It also highlights the need to monitor the practice of removing fuses for maintenance and plant modifications, which has been identified as a cause of fatigue of the fuse holder clips.

4.12   Information Notice 83-50, "Failures of Class 1E Safety-Related Switchgear Circuit Breakers to Close on Demand"

This notice presents the results of an NRC study on safety-related switchgear circuit breaker failures. A review of past operating experience found a number of failures to close on demand. One of the causes attributed to these failures was blown control circuit fuses, which resulted in the malfunctioning of the circuit breaker's closing control circuitry. These results are significant in terms of plant safety since switchgear circuit breakers are used to operate numerous safety-related equipment in the plant.

4.13   NRC Weekly Information Report - Week Ending July 4, 1997

This report documents an event in June 1997 at the Oconee Station in which emergency power from the Keowee Hydro Station was called for during a test. The Keowee Hydro Station acts as the emergency power supply for the plant instead of diesel generators. Keowee Unit 2 responded as expected, however, the Unit 1 generator failed to develop a generator field due a blown fuse in the field flashing circuit, and was unable to provide power to the Oconee emergency bus. The cause of the blown fuse appeared to be age-related. The fuse was replaced and power was supplied successfully. This event is significant since the fuse failure caused a loss of redundancy in the emergency power supply to the plant.

4.14   Evaluation of Events from NRC Documents

The fuse related events discussed above demonstrate that age-related degradation of fuses does occur in nuclear plant applications and does lead to safety significant plant effects. Generic fuse problems, such as failure of brass ferrules and cracking of fuse elements in a specific brand of fuses have been identified, and actions have been taken by the manufacturers to address these problems. However, there still exist age-related issues that continue to cause failures and need to be addressed, such as the mechanical failure of the indicating type fuses. Fuse issues related to human error, as well as plant procedures and practices also exist that can lead to equipment failure. Examples of these include the installation of incorrect fuses in the Westinghouse 7300 circuit cards and the practice of removal and reinsertion of fuses to secure circuits, equipment, and systems for maintenance and testing.

Since fuses are used throughout nuclear power stations for both safety-related and non safety-related applications, their failure can have a significant impact on plant performance. Typical plant effects observed from this review are challenges to safety systems, reactor trips, or loss of redundancy of safety trains. Clearly, managing age-related failures of fuses would have a positive effect on the safety performance of a plant. An awareness of the potential fuse failure problems and aging mechanisms is one important step in mitigating the consequences of such fuse failures.

Table 3 summarizes the fuse events identified from the review of NRC documents.

Table 3 Summary of fuse events identified from a review of NRC documents

Document	Event Summary	Fuse Type	Failure Mechanism	Failure Cause	Failure Mode	Aging Related?
1. Part 21 Report Concerning Cracked Fuse Ferrules - Accession #9612170255	Cracking of brass ferrules caused potential catastrophic fuse failure.	Fuses using brass ferrules manufactured by: • Gould-Shawmut, • Bussmann, and • CEFCO	Stress corrosion cracking	Stress buildup plus corrosive flux residue from the manufacturing process	Loss of Continuity	Yes
2. Part 21 Report Concerning Cracking of Fuse Elements - Accession #9604030315	Cracking of fuse element caused fuses to fail.	Gould-Shawmut 10-15 amp fuses	Embrittlement	Corrosive flux residues used in the soldering process	Fuse Opens Spuriously	Yes
3. Information Notice 96-23, "Fires in Emergency Diesel Generator Exciters During Operation Following Undetected Fuse Blowing"	Fuses with no blown fuse indicator failed and went undetected until they led to a fire in the emergency diesel generator exciter.	Various	Various	No blown fuse indicator	Fuse Opens Spuriously	No
4. Information Notice 93-87, "Fuse Problems with Westinghouse 7300 Printed Circuit Cards"	Use of incorrectly sized fuses caused failure of other components on circuit card.	Various	Various	Human error	Fuse Opens Early or Fuse Opens Late	No
5. Information Notice 91-78, "Status Indication of Control Power for Circuit Breakers used in Safety-Related Applications"	Deformed fuse holder fingers caused a containment spray pump to fail to start	Various	Fatigue, deformation, or degradation	Repeated handling to remove and replace fuses	Loss of Continuity	Yes

Table 1 Summary of fuse events identified from a review of NRC documents (continued)

Document	Event Summary	Fuse Type	Failure Mechanism	Failure Cause	Failure Mode	Aging Related?
6. Information Notice 91-51, "Inadequate Fuse Control Programs"	Inadequate fuse control identified as a potential cause of safety-related equipment failure	Various	Various	Human Error	Various	No
7. Information Notice No. 87-42, "Diesel Generator Fuse Contacts"	Fuse holder contact degradation caused over-temperature condition, which led to cable insulation failure.	Various	Misalignment or corrosion of fuse holder contacts	Exposure to various operating stressors	High Resistance	Yes
8. Information Notice 87-62, "Mechanical Failure of Indicating-Type Fuses"	Fuses failed as a result of mechanical failure of the fuse element.	Indicating type fuses manufactured by: • Bussmann, and • Littelfuse	Degradation of solder joint, creep, or fatigue of fuse element due to spring load	Cold solder joint plus exposure to operating stressors	Fuse Opens Spuriously	Yes
9. Information Notice 87-24, "Operational Experience Involving Losses of Electrical Inverters"	Fuse failures identified as one potential cause of inverter failures	Various	Fuse element degradation due to oxidation or corrosion	Exposure to high temperature or humidity conditions	Various	Potentially
10. Information Notice 86-87, "Loss of Offsite Power Upon an Automatic Bus Transfer"	Fuse failure caused reactor trip	Various	Loosening of fuse holder	Exposure to repeated handling or vibration	High Resistance	Potentially
11. Information Notice 85-51, "Inadvertent Loss or Improper Actuation of Safety-related Equipment	Fuse failures identified as a potential cause of safety-related equipment failure	Various	Various	Various	Various	Potentially

Table 1 Summary of fuse events identified from a review of NRC documents (continued)

Document	Event Summary	Fuse Type	Failure Mechanism	Failure Cause	Failure Mode	Aging Related?
12. Information Notice 83-50, "Failures of Class 1E Safety-Related Switchgear Circuit Breakers to Close on Demand"	Fuse failures identified as one potential cause of switchgear circuit breaker failures	Control circuit fuses	Various	Various	Various	Potentially
13. NRC Weekly Information Report - Week Ending July 4, 1997	Fuse failure identified as cause of loss of emergency power	Generator exciter fuse	Age-related degradation of fuse element	Exposure to operating stressors	Fuse Opens Spuriously	Potentially

5.0   Review of Operating Experience

To gain insights into potential age-related failures being experienced by fuses in nuclear power plant applications, a review of past operating experience for both safety-related and non safety-related fuses was performed. While the focus of this study is on safety-related fuses, the evaluation of non safety-related fuses was included since they are constructed of the same materials and are exposed to the same stressors as safety-related fuses. This also allowed an evaluation of the effects of fuse failures in non-safety-related systems on overall plant safety. Thus, additional data is provided, which improves the accuracy of the evaluation.

Operating experience data were obtained from the Licensee Event Report (LER) database, the Nuclear Plant Reliability Data System (NPRDS) database, and the Equipment Performance Information Exchange System (EPIX) database. LERs are submitted to the NRC by commercial nuclear power plants (licensees) in response to reporting requirements as defined in the Code of Federal Regulations. Their scope is limited to safety-related equipment and systems in the plant, and the reports include equipment failures as well as administrative deficiencies.

The NPRDS is a national database of events related to both safety- and non safety-related equipment and systems in nuclear power plants. It includes events from the early 1980's to 1996, and is focused on collecting information related specifically to equipment failures, making this database an excellent source of information for evaluating the performance of aged equipment. Licensees submit reports to the NPRDS on a voluntary basis, and not all plants report to this database. The information reported typically includes a description of the event, the component involved, and the circumstances of the event. While a standardized reporting format exists, the content and detail of the reports can vary considerably depending on the organization submitting the report. In 1997 the EPIX database replaced the NPRDS.

The results of these data reviews are discussed in the following sections.

5.1   LER Operating Experience

The LER database was searched for age-related fuse events over the 20-year period from 1981 to 2001 using a field search for "fuse" failures caused by "wearout/age/end of life." This was supplemented by keyword searches covering the 10-year period from 1991 to 2001 to ensure that the majority of the age-related fuse events were captured. The keyword searches used were "fuse" NEAR "aging" ; "fuse*" NEAR "normal wear" ; "fuse*" NEAR "wear out" ; "fuse*" NEAR "degraded" ;and "fuse" NEAR "failed." Out of 114 units in operation during the reporting period searched, the following report statistics were identified from the search results:

- A total of 42 units reported at least one event involving fuses
- A total of 630 events were identified and reviewed involving fuses
- A total of 55 age-related fuse failures were identified

It is significant to note that of the 114 nuclear power plants operating during the period from 1981 to 2001, only 42 (37%) experienced fuse failures requiring a LER. Considering the large number of fuses in operation at each plant, this suggests that failures of safety-related fuses

23

are relatively infrequent. The distribution of age-related failures reported per unit is shown in Figure 4. For the units reporting, the fuse failures were relatively evenly distributed among the units, with no one unit having a majority of the failures. This indicates there is no specific age-related fuse problem at any one plant. In most cases, only one fuse failure per unit was reported.

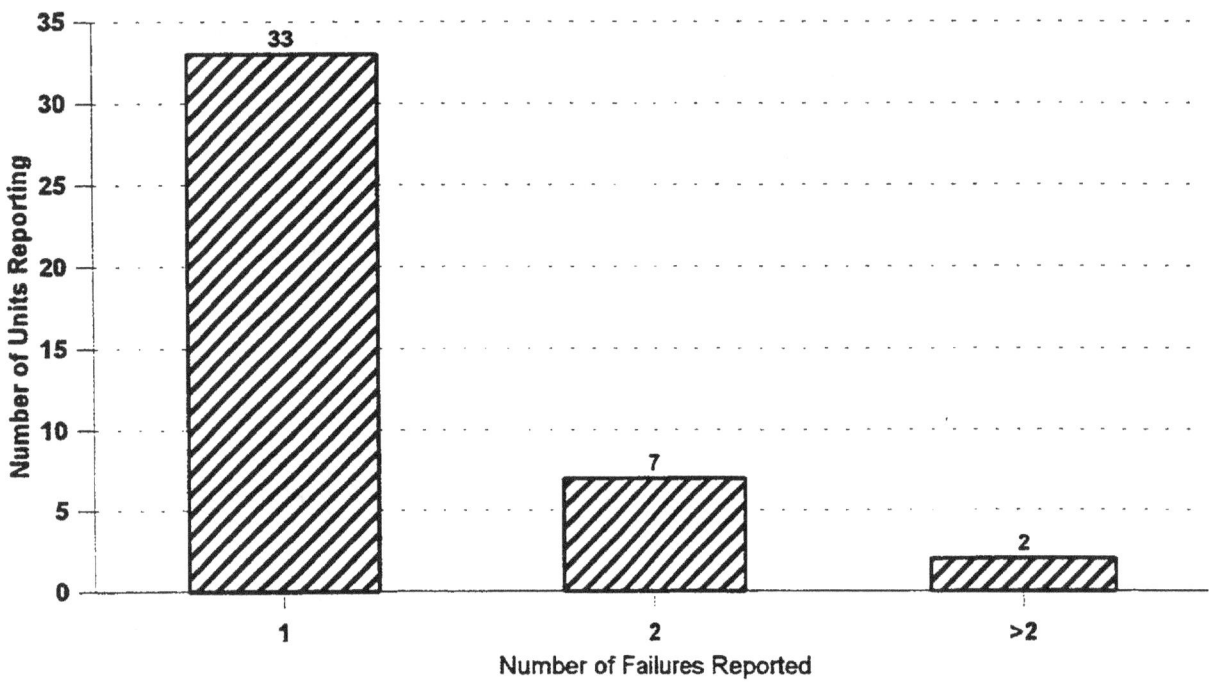

Figure 4 Distribution of age-related fuse failures per unit from LERs

The age-related fuse failures were analyzed to identify the predominant aging characteristics of the failures experienced. Figure 5 shows the distribution of failure modes identified. As shown, the predominant failure mode experienced (82%) was "FUSE OPENS SPURIOUSLY," which indicates that the fusible element opened and caused an open circuit when it should not have. The failure mode "HIGH RESISTANCE/LOSS OF CONTINUITY" represented 9% of the failures. These failures typically involved problems with the fuse being loose in the fuse holder due to weakened clips and failing to complete the circuit, and/or degradation of the electrical contact surfaces. The remaining events (9%) involved the failure mode "INTERMITTENT OPERATION" of the fuse, which were typically due to loose fuse holder clips or a broken solder joint between the fuse element and end cap. No events were identified involving any of the other potential fuse failure modes identified previously.

The predominant failure mechanism (80%) was found to be "FATIGUE/ DEGRADATION" of the fusible link, which led to failure of the fuse. Fatigue is typically due to degradation over a period of time from exposure to elevated temperature, voltage transients, or short duration over-current conditions. It can lead to weakening of the fuse element, or a reduction in cross section, which reduces its current carrying capacity. Other less frequent failure mechanisms found were "WEAR/FATIGUE" of the fuse holder clips (9%) or "DEGRADATION" of the end

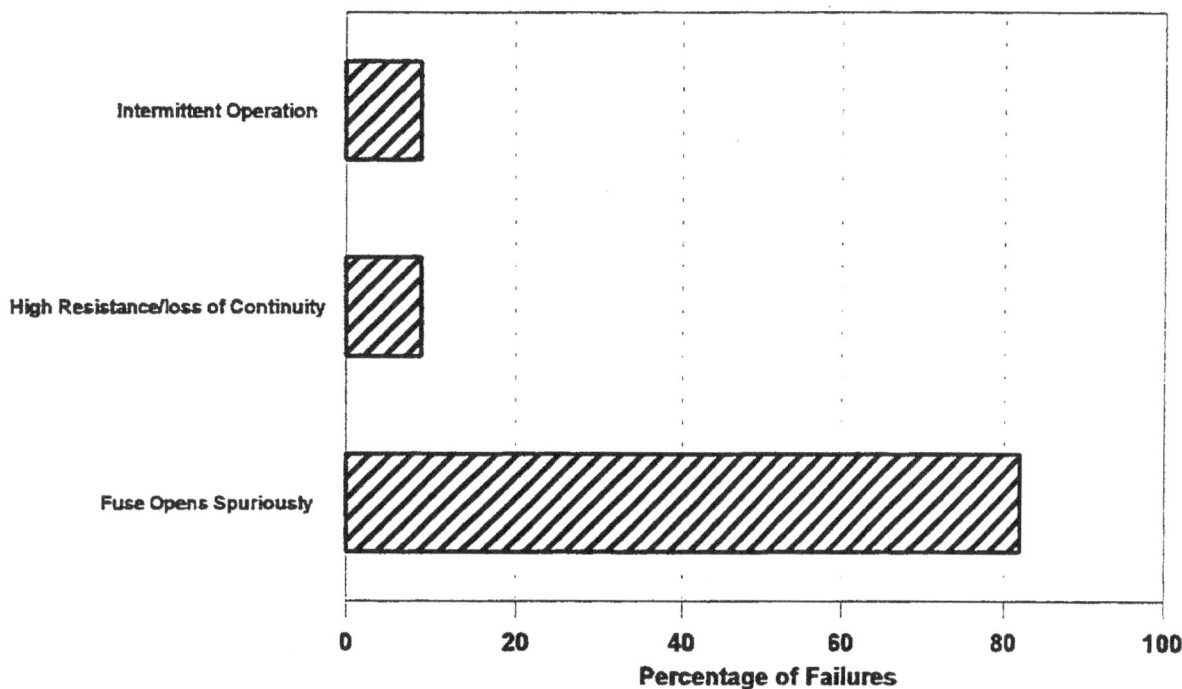

Figure 5 Distribution of fuse failure modes from LERs

cap to fuse element solder joint (7%). Fatigue of the fuse holder clips can typically occur due to repeated insertion and removal of the fuses, such as during maintenance or surveillance testing. Degradation of the solder joint was found to occur in several cases due to cold solder joints that degraded during subsequent operation. The remaining failures were due to loosening/degradation of the wiring connections to the fuse holder, and loosening/ wear of the end caps on the fuse itself. The percentage of failures related to each of the failure mechanisms is shown in Figure 6.

The subcomponent most frequently failed was found to be the fusible link (80%), which is consistent with the predominant failure mechanism identified. Typically, the LERs report fuse failures in which the fusible link blows and the cause cannot be attributed to the circuit. In these cases, the failure is attributed to aging of the fuse that resulted in fatigue of the fusible link due to operation over long periods of time. For purposes of this analysis, in cases where the report states that the fuse had "blown" and no information is given as to the subcomponent that failed, it is assumed that fatigue or degradation of the fuse element occurred. This assumption was judged to be justifiable for several reasons; 1) the fuse element is designed to be the weakest link in the fuse and, therefore, the most susceptible to failure under normal circumstances, 2) it would be more likely for failure of a subcomponent other than the element to be mentioned in a failure report, and 3) it is common industry terminology to refer to failure of a fuse due to melting of the element as a "blown fuse." Other subcomponents that failed less frequently were the fuse holder clips (9%), the connection of the fusible link to the end caps (7%), the fuse holder wiring lugs (2%), and the fuse blades/ferrules (2%).

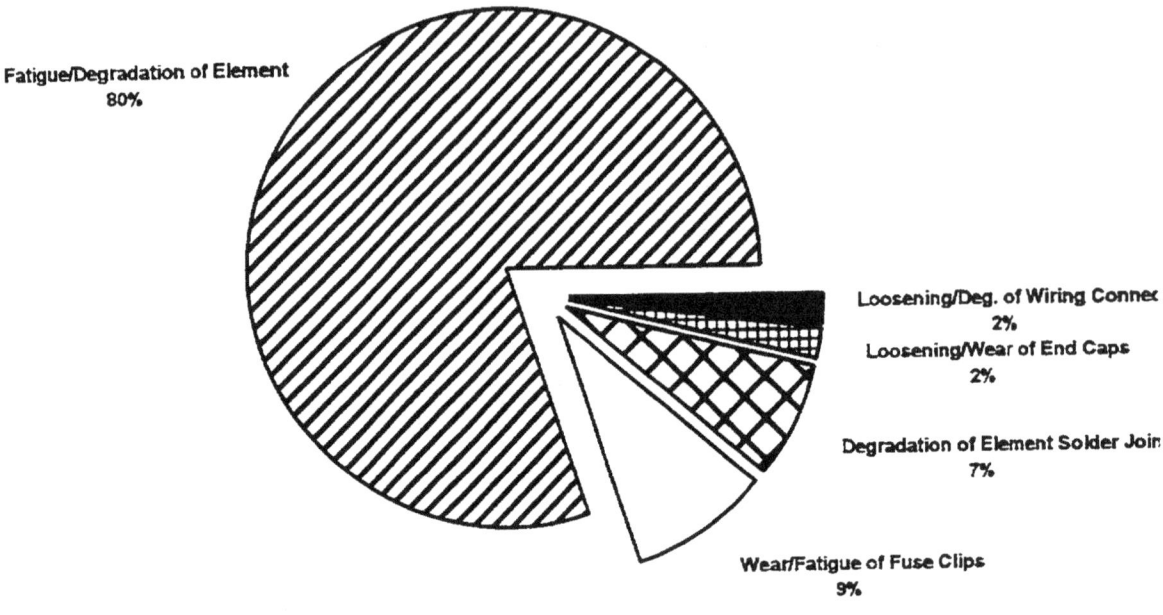

Figure 6 Fuse failure mechanisms from LERs

The various effects of the fuse failures on plant performance are shown in Figure 7. The predominant effect (42%) was found to be a challenge to a safety or backup system. This is expected from this database since any actuation of an engineered safety feature is reportable as an LER. Typical examples of these events are the following:

- LER 29689005: At Browns Ferry, Unit 3, in October 1989, two radiation monitors for the reactor and refueling zone experienced a loss of power and tripped upscale. This resulted in the automatic actuation of the logic for several engineered safety features, including control room emergency ventilation, standby gas treatment, refuel zone ventilation isolation, and primary containment isolation. The cause of the power failure was attributed to the failure of a fuse on the voltage regulator board in the radiation monitor power supply. Testing of the power supply indicated there were no faulty components that could have caused the fuse failure. Therefore, the cause of the fuse failure was attributed to fatigue.

- LER 31192001: At the Salem Generating Station, Unit 2, in January 1992, a vital bus tripped during a transfer of the bus from one station power transformer to another. The infeed breaker for the latter bus did not close due to failure of a cartridge fuse in the potential transformer secondary, which supplies power to the metering circuit. The failure of the vital bus resulted in the automatic start and loading of the emergency diesel generator, which is an Engineered Safety Feature. No problem was found with the circuit and the fuse was replaced. This was considered to be an isolated occurrence.

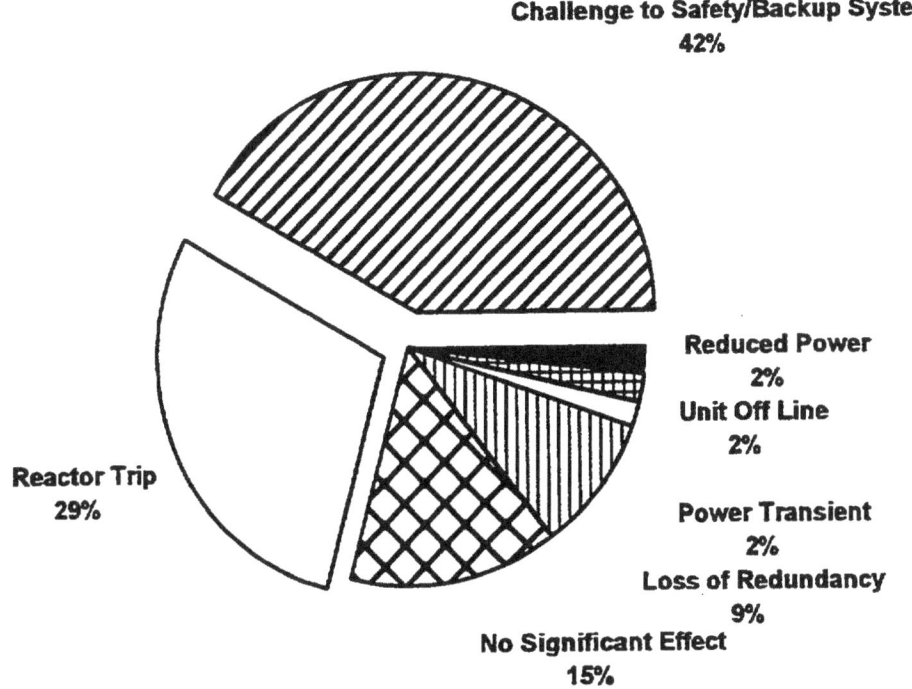

Figure 7 Effect of fuse failures on plant performance from LERs

- LER 32191016: At Plant Hatch, Unit 1, in September 1991, a control power fuse for a containment isolation valve control logic circuit failed resulting in spurious closure of a primary containment isolation valve. This in turn tripped the reactor water cleanup pump. No problems were found with the logic circuit. The failure was considered a random event, and the fuse was replaced.

The second most frequent effect of fuse failure was a reactor trip. These events also involve actuation of engineered safety features, however, they are more severe and ultimately include a trip of the reactor. Typical examples of these events are the following:

- LER 24788019: At Indian Point, Unit 2, in November 1988, a reactor trip occurred due to low steam generator level. This was caused by the unintended closure of a feedwater regulating valve. The valve failed closed when a control power fuse in its control circuit failed. The circuit was tested and found to be acceptable, therefore, the fuse failure was attributed to its age. The fuse was a Bussmann model B34-24, 3 amp, 250 Vdc.

- LER 31396007: At Arkansas Nuclear One, Unit 1, in September 1996 an automatic reactor trip was initiated when a non-vital 6.9 kV electrical bus providing power to two reactor coolant pump motors sensed an under voltage condition and opened, causing loss of the pumps. The cause was attributed to mechanical failure of a fuse in the bus protection circuitry, which was interpreted as an under-voltage condition on the bus. A defective solder joint between the element and end cap degraded while in service and resulted in interruption of continuity. The fuse, a model OT10 manufactured by Gould

27

Shawmut, was approximately 27 years old at the time of failure. The fuse was constructed using a non-laminated paper cartridge with metal end caps crimped at each end. The fuse element is contained within the cartridge, soldered to the end caps, and protected with an inert silica powder filler material.

- LER 53094007: At Palo Verde, Unit 3, in August 1994, a reactor trip occurred when steam generator water level reached the trip setpoint for high level. This was caused by an unintended increase in main feedwater flow due to failure of the feedwater flow controller. This was attributed to a failure of the flow controller master controller power fuse.

To examine the effect of fuse failures further, the data were analyzed to determine the application of the fuses that failed. The predominant application was found to be for "LOGIC CONTROLLERS." This application typically involves providing control power to a logic controller, which controls the operation of a piece of equipment. Examples include the logic controllers for actuation of engineered safety features, such as containment isolation valves and control rod drive mechanisms. Logic controllers are also used for automatic control of various system processes, such as feedwater flow control. Other applications for the fuses that failed include power supplies to radiation monitors and transformer circuit breakers. Figure 8 shows the distribution of fuse applications from the LER failure reports.

The LER data were also sorted to identify the predominant method of detecting fuse failures, and the results are shown in Figure 9. The predominant number of fuse failures (65%) were detected by operational abnormalities, such as the spurious start of a piece of equipment or the unintended actuation of an engineered safety feature. This was followed by audiovisual alarms (18%) and surveillance testing (15%). The fact that most fuse failures are not detected until an operational abnormality occurs suggests that improved methods of monitoring the condition of fuses would be beneficial.

5.2   NPRDS Operating Experience

The NPRDS database includes operating experience from the early 1970's through 1997. The database was found to contain a total of 3,262 fuse-related failure reports and 1,145 of these were reviewed to provide operating experience data for this study. The events reviewed generally included all fuse-related failures reported to NPRDS since 1996, supplemented by keyword searches from 1980 to 1997 specifically for "control power fuse," "fuse holder," "fuse normal wear," "fuse end of life," and "fuse wear out." Finally, all fuse failures reported from 1990 through 1997 for a group of 22 nuclear units were reviewed, to ensure a thorough sampling of the full range of fuse failures. Out of 114 units operating during the reporting period covered, the following statistics were obtained from the NPRDS operating experience review:

- A total of 94 units reported at least one event involving fuses
- A total of 1,145 events were identified and reviewed involving fuses
- A total of 354 age-related fuse failures were identified

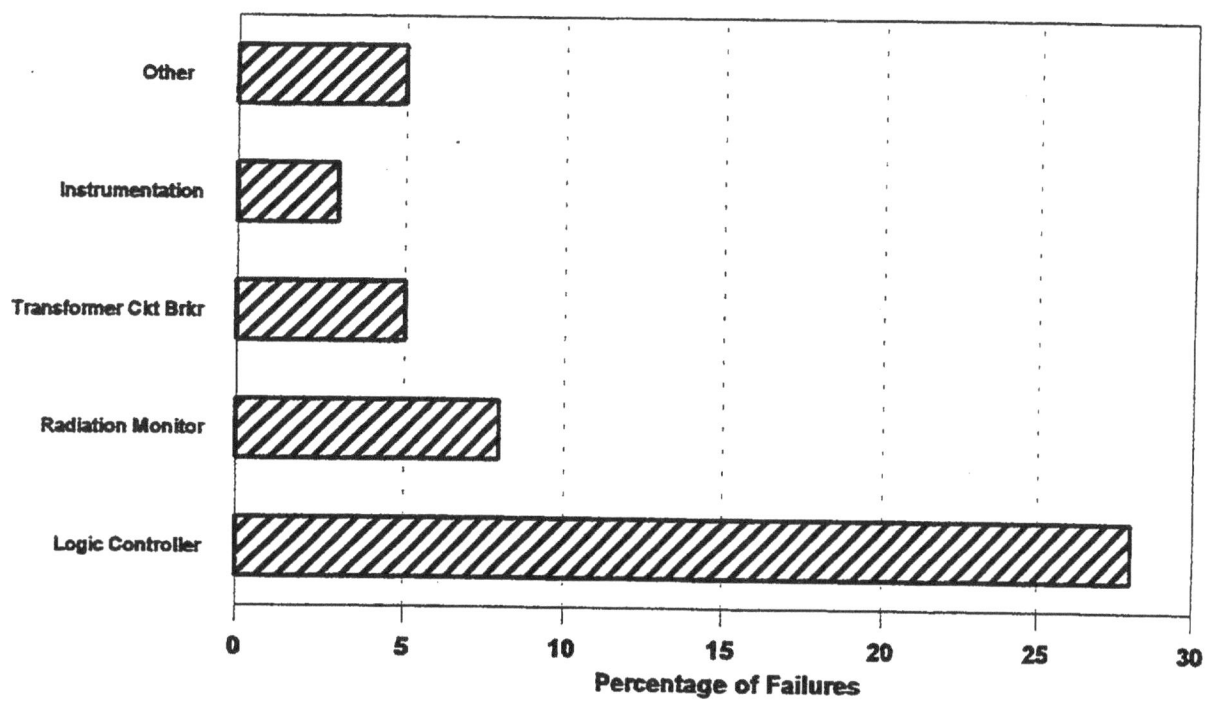

Figure 8 Distribution of applications for failed fuses from LERs

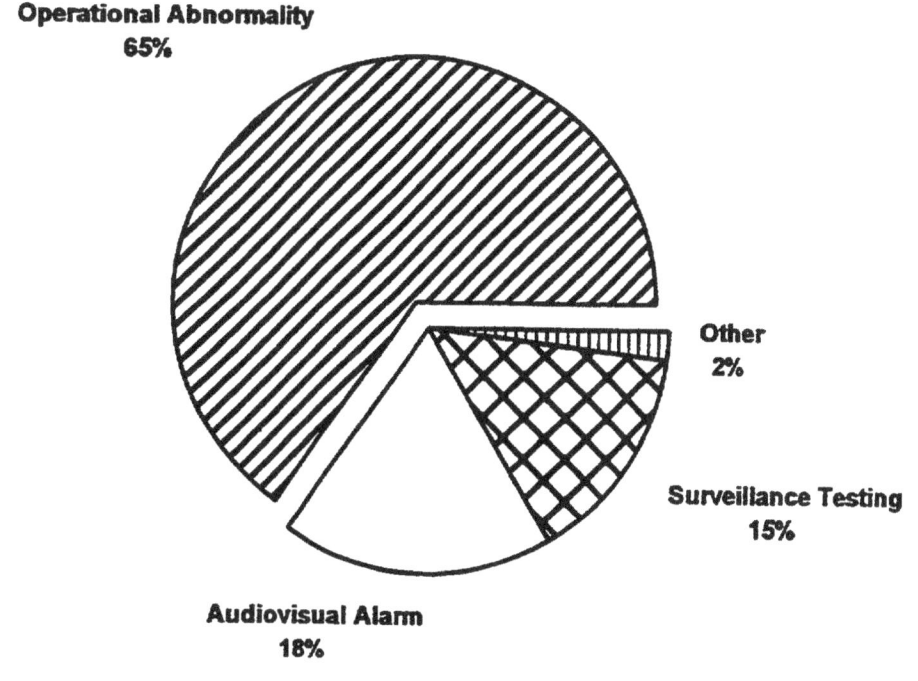

Figure 9 Fuse failure detection methods from LERs

Considering the thousands of fuses of all sizes and types that are in operation at each of the 114 nuclear power units that operated during this period, the number of age-related events among the failures reported to the NPRDS was relatively low. The failure reports reviewed for this program represented 94 nuclear units. The distribution of age-related failures reported per nuclear unit is shown in Figure 10.

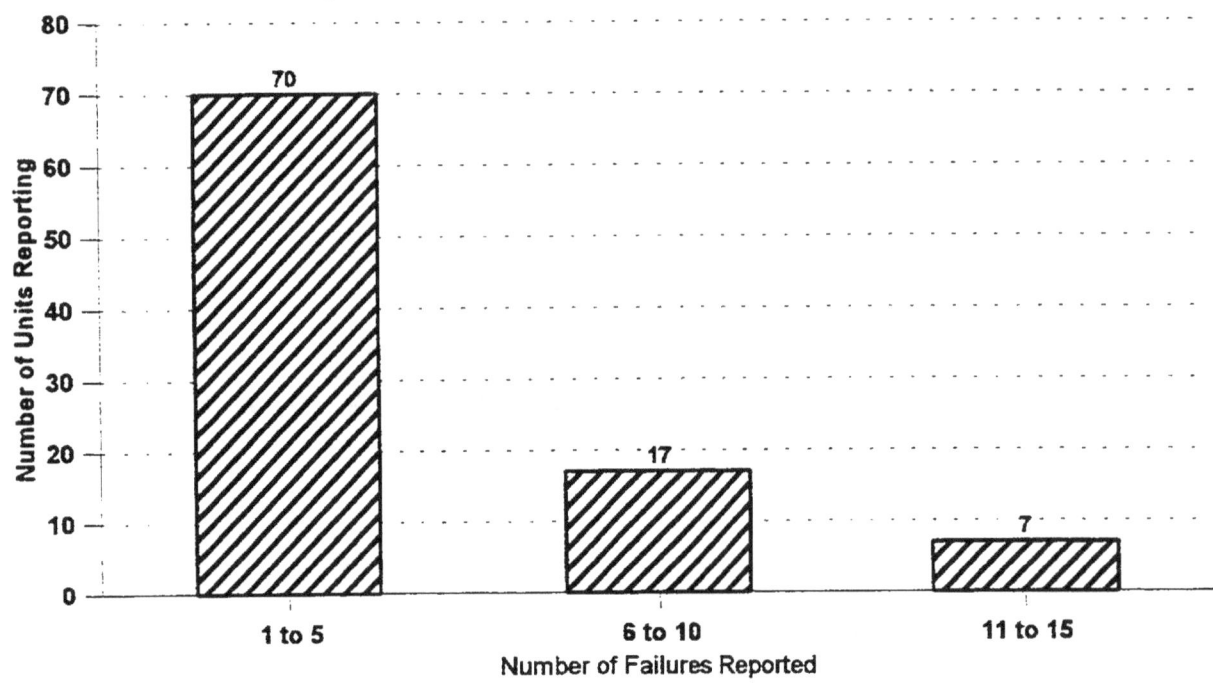

Figure 10 Distribution of age-related fuse failures per unit from NPRDS

The fuse failures identified as age-related were analyzed to determine the predominant aging characteristics of the failures experienced. Figure 11 shows the distribution of failure modes encountered during the review. As shown, the predominant failure mode experienced (64%) was "FUSE OPENS SPURIOUSLY," which indicates that the fusible element opened and caused an open circuit when it should not have. The failure mode "HIGH RESISTANCE/LOSS OF CONTINUITY," which represented 33% of the age-related failures reported, was significant in that it was a failure mode of the fuse holder. These failures typically involved: problems with the fuse being loose in the fuse holder due to weakened, broken, or misaligned clips; degraded or open electrical connection between fuse and holder; degraded or open electrical connection between the fuse holder and its wiring terminations; broken fuse holder parts; and failing to complete the circuit.

A small number of events (2.5%) involved the failure mode "INTERMITTENT OPERATION" of the fuse, which were typically caused by loose fuse holder clips or faulty fuse holder wiring terminations. Finally, less than 1% of the events involved the "GROUND FAULT" failure mode, which was usually caused by dirt or contamination on a high voltage fuse/fuse holder assembly faulting to ground. It should be noted that latter three failure modes, representing more than 35% of the age-related events reviewed, involved failures of the fuse holder assembly and its terminal connections. No events were reviewed involving any of the other potential fuse failure modes identified previously.

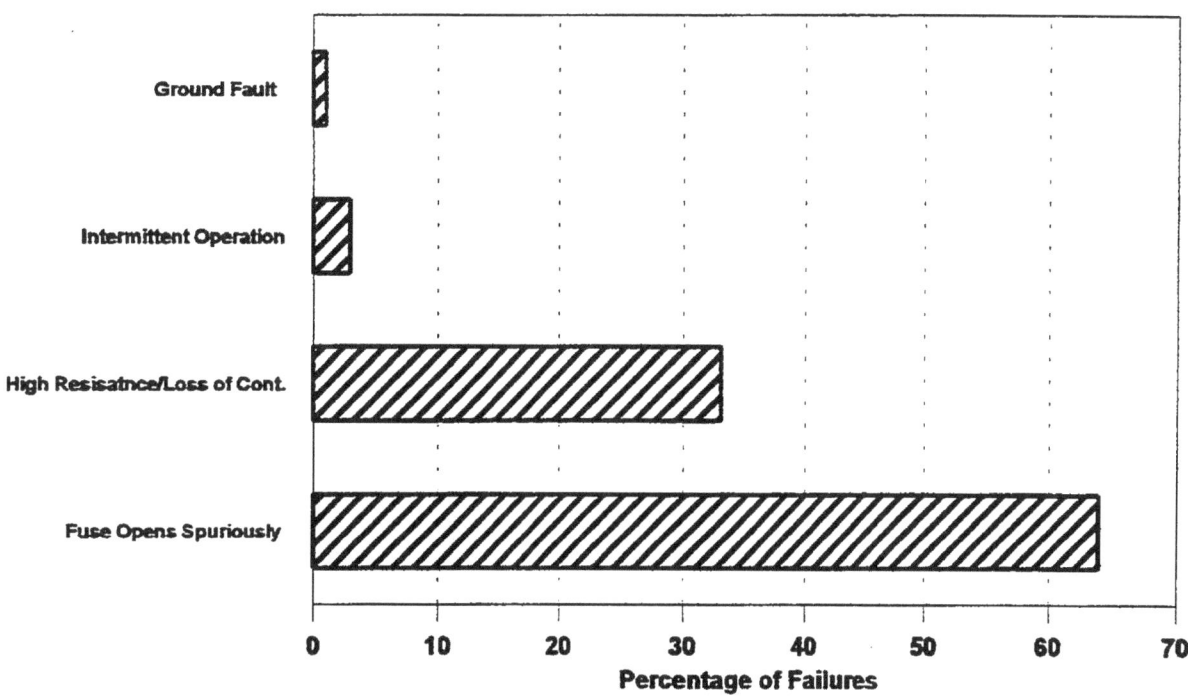

Figure 11 Distribution of fuse failure modes from NPRDS

The predominant failure mechanism (61%) was found to be "FATIGUE/DEGRADATION OF ELEMENT," i.e., of the fusible element or link, which led to unexpected failure of the fuse. This is to be distinguished from the normal opening of the fusible link when the fuse is exposed to overcurrent conditions for a prescribed time. Fatigue is typically due to the degradation of the metallic fuse element over time as a result of exposure to elevated temperature, voltage transients, or short duration overcurrent conditions. It can lead to weakening of the fuse element, or a reduction in cross section, which reduces its current carrying capacity.

A second important failure mechanism was "WEAR/FATIGUE OF FUSE CLIPS" (21%). Fatigue of the fuse holder clips can typically occur due to high temperature, mechanical stress, and repeated insertion and removal of the fuses, such as during maintenance or surveillance testing. Other less frequently cited failure mechanisms found were "CORROSION/DEGRADATION OF CONTACT SURFACES" (6%), loose, broken, or degraded soldered (3%) or bolted/screwed (4%) wiring connections, and "LOOSENING/WEAR OF END CAPS" (1%). The remainder of the failure mechanisms (approximately 4%) included a variety of mechanisms, such as, contamination, moisture intrusion, fatigue or breakage of fuse holder base, or unknown. The percentage of failures related to each of the failure mechanisms is shown in Figure 12.

31

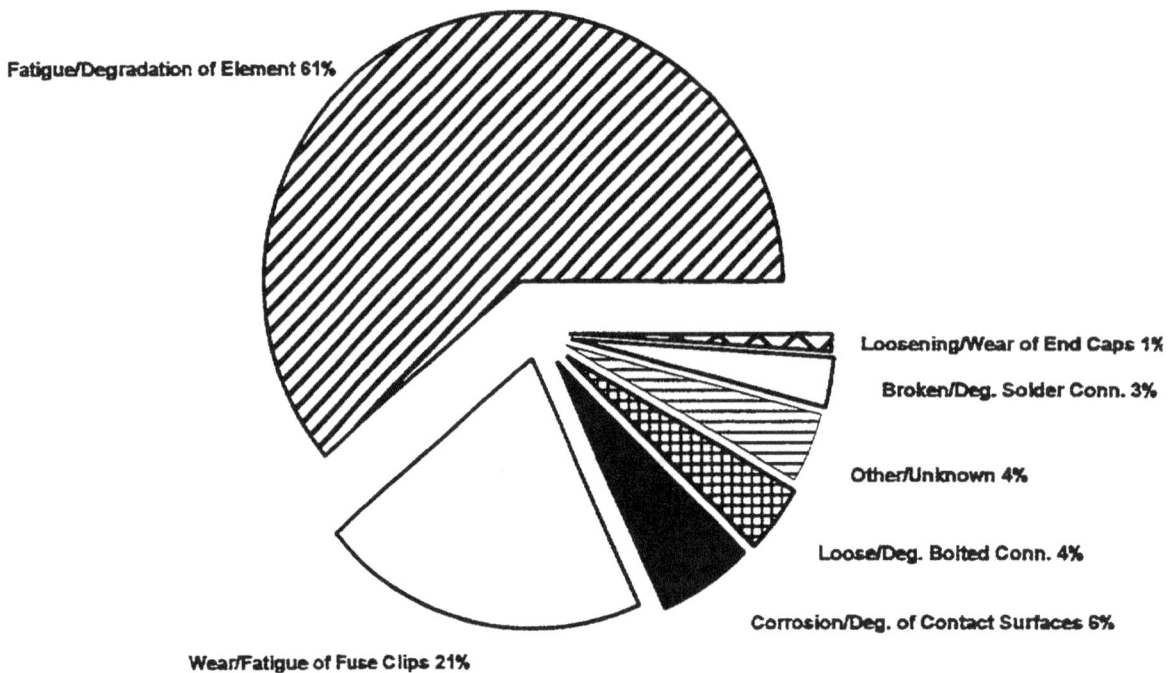

Fatigue/Degradation of Element 61%

Loosening/Wear of End Caps 1%

Broken/Deg. Solder Conn. 3%

Other/Unknown 4%

Loose/Deg. Bolted Conn. 4%

Corrosion/Deg. of Contact Surfaces 6%

Wear/Fatigue of Fuse Clips 21%

Figure 12 Fuse failure mechanisms from NPRDS

The subcomponent most frequently reported to NPRDS as failed was the fusible link (60%), which is consistent with the predominant failure mechanism identified. Detailed root cause analyses were not normally found in NPRDS reports. However, if a fuse link opens spuriously when no related maintenance or operating activities are taking place, and no other circuit problems are revealed during troubleshooting, the event is normally reported as a wear out failure of the fuse link. In these cases, the failure is attributed to aging of the fuse that resulted in fatigue of the fusible link due to operation over long periods of time. Another important group of subcomponent failures is related to the fuse holder assembly and its wiring connections to the circuit that is being protected. Fuse holder-related subcomponent failures reported to NPRDS included fuse holder clips (15%), wiring connectors (7%) (including lugs, screws, pressure terminals, solder joints), general fuse holder assembly (11%) (including fuse holder base, clip mounting fasteners, base mounting fasteners), and blades/ferrules (6%). This is consistent with the fuse holder-related failure modes described above.

The effects of fuse failures on plant performance are shown on Figure 13. In contrast to the LER data base, which by its nature involves events that have the most serious effects on nuclear plant operation, the NPRDS is a general equipment and component failure data base. Hence, most of the fuse failures reported to NPRDS resulted in "NO SIGNIFICANT EFFECT" (95%) on plant operation. Among the fuse failures reviewed, 3% resulted in "REDUCED POWER" operations, 1% resulted in a "REACTOR TRIP", and 1% resulted in the taking the "UNIT OFF LINE."

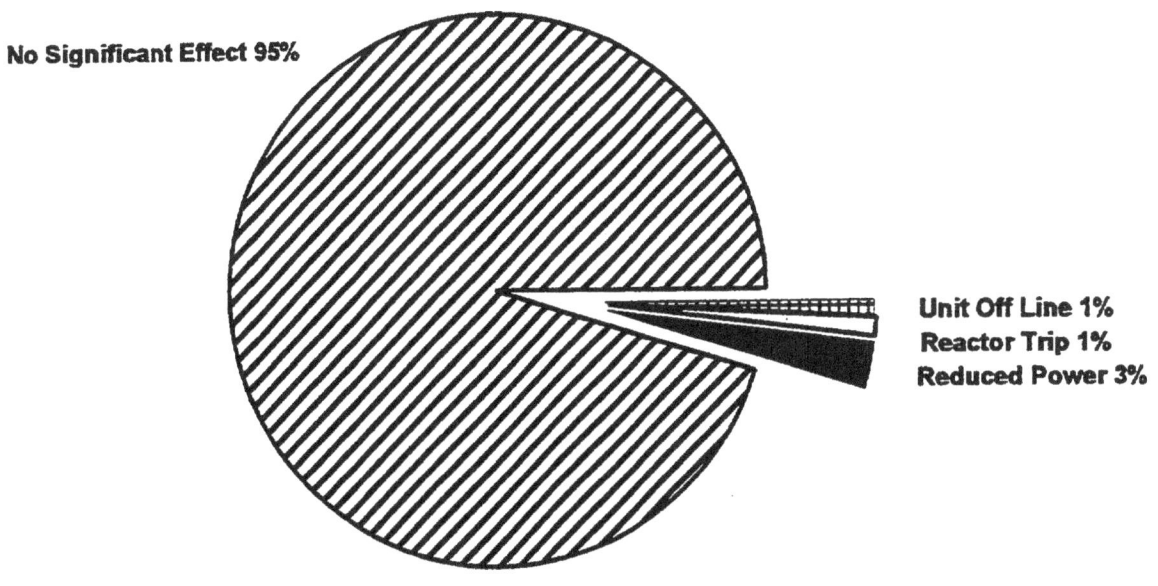

No Significant Effect 95%

Unit Off Line 1%
Reactor Trip 1%
Reduced Power 3%

Figure 13 Effect of fuse failures on plant performance from NPRDS

Another measure of the significance of fuse failures can be found by looking at the effect of a fuse failure upon the system in which it is located. This information is included as part of the NPRDS failure report and is shown on Figure 14 for fuse failures. More than three-fourths of the failures result in either "DEGRADED TRAIN/CHANNEL" (38%) or the "LOSS OF ONE OR MORE TRAINS/CHANNEL FUNCTIONS" (40%). Another 5% of the fuse failures caused "DEGRADED SYSTEM OPERATION," and the "LOSS OF ONE OR MORE SYSTEM FUNCTIONS" was reported in 1% of the failures. Finally, in 16% of the fuse failures reported to NPRDS, the plants indicated that there was "NO EFFECT" at all on the system in which the problem was found.

Fuse failures, once they are found, are rapidly and easily corrected, often requiring nothing more than troubleshooting and replacement of a blown fuse. The NPRDS data showed that more than 34% of the fuse failures were corrected on the same day that they were discovered and another 22% were corrected by the next day. Within 7 days, nearly 85% of the reported fuse failures were corrected and, after two weeks, almost 90% of the fuse-related problems had been repaired and their systems returned to service.

Another important piece of information found in the NPRDS data is the application in which the failed fuse assembly was being used. These data can provide insight into how fuse failures may affect plant operation and safety, and suggest approaches that may be taken to mitigate the effects of fuse failures. The fuse application data are presented in Figure 15. The most common application involved control power fuses for motor operated valves and dampers (23%) or solenoid operated valves (5%). Control power fuses for electric drive motors for pumps, fans, and blowers were involved in 18% of the failure reports. Also significant were

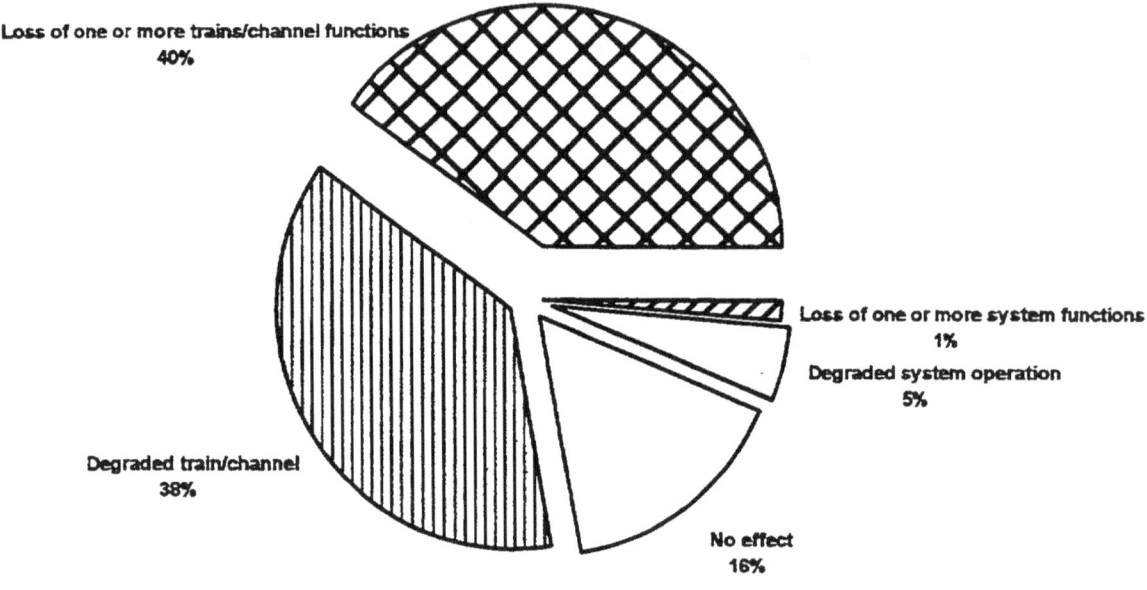

Figure 14 Effect of fuse failures on system performance from NPRDS

electronic circuit card fuses for process control circuits and systems (16%) and closely related were instrument power supply fuses (14%). Finally, a fairly large number of fuse failures occurred in large power supply/rectifier applications (4%) and in battery chargers/inverters/ uninterruptible power supplies (7%). The latter application was the subject of an earlier BNL study on the operating experience and aging-seismic assessment of battery chargers and inverters (reported in NUREG/CR-4564 and NUREG/CR-5051), which included an examination of the fuse failures experienced in this equipment that were potentially the result of elevated operating temperatures.

The remainder of the applications found included: fuses in the emergency diesel generator logic, instrumentation and control circuits, and output circuit breaker control power circuits; logic and control circuit fuses for steam turbines; control and instrumentation fuses for M-G sets; power feeder circuit breaker control power fuses; and potential transformer primary and secondary circuit fuses.

The NPRDS data were also sorted to identify the predominant methods for detection of fuse failures. This information can sometimes show whether the methods and techniques being used to discover degraded or failed fuses are well-matched with the predominant failure mechanisms encountered for this component. The NPRDS failure detection data are shown on Figure 16. The greatest number of fuse failures were detected during "MAINTENANCE" and "TESTING" activities (39%) and by "OBSERVATION" (15%) by plant operating and maintenance personnel. This is significant because it reflects proactive efforts on the part of licensees to find these failures before they can cause more serious problems. Another 25% of the NPRDS failures were detected by "OPERATIONAL ABNORMALITIES." This category would include failures to operate when required, off-normal performance, loss of position

34

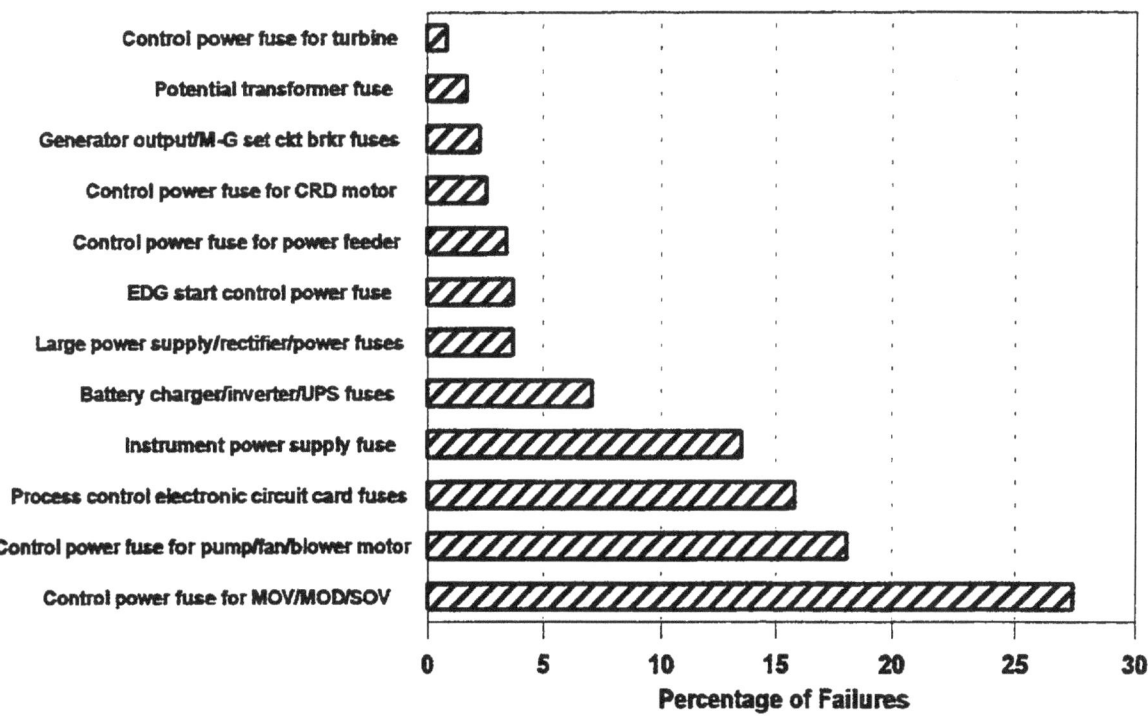

Figure 15 Distribution of applications for failed fuses reported to NPRDS

indication, or loss of control power indication. Finally, "AUDIOVISUAL ALARMS" accounted for the detection of 21% of the fuse failures. It should be noted that several NPRDS reports

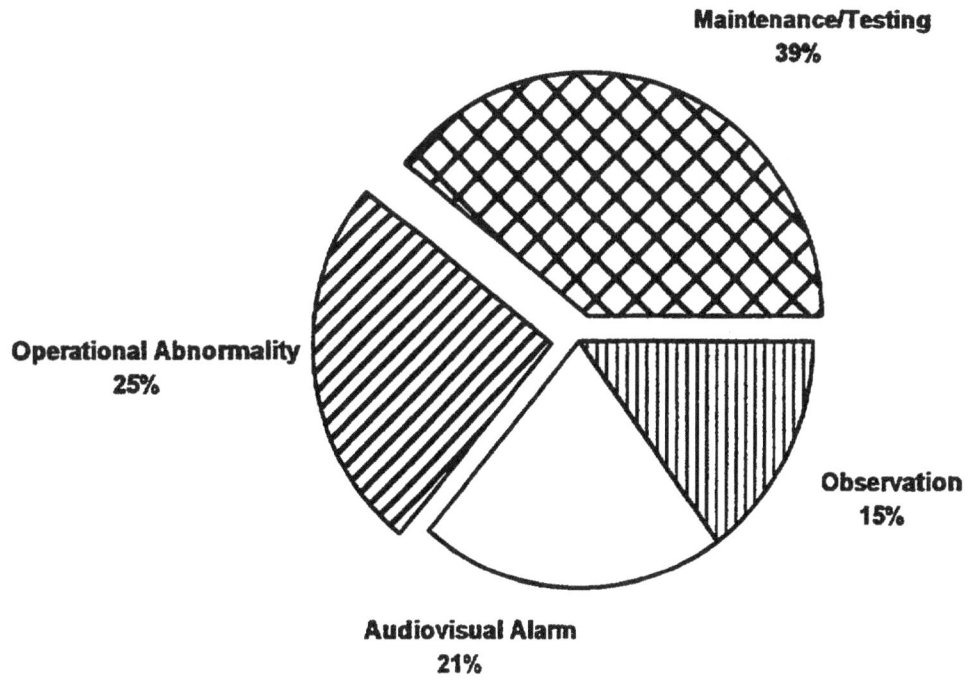

Figure 16 Fuse failure detection methods from NPRDS

mentioned that imaging infrared thermography surveys were used to identify incipient fuse and fuse holder failures. Licensees are taking advantage of this technology to detect hot spots that could indicate potential fuse assembly failures.

5.3   EPIX Operating Experience

The EPIX database replaced the NPRDS starting in 1997 and includes operating experience from the years 1997 to 2001. While the information to be included in EPIX is similar to NPRDS, it was found to be less detailed and less comprehensive. Therefore, not all of the same analyses could be performed; for example, the distribution of applications for the fuse failures reported to EPIX could not be determined.

A total of 165 fuse-related failures were extracted from this data. Of these, 37 failures were related to aging effects. Out of the 106 units operating during the reporting period covered, the following statistics were obtained from the search results:

- A total of 27 units reported at least one event involving fuses
- A total of 165 events were identified and reviewed involving fuses
- A total of 37 age-related fuse failures were identified

Again, it is significant to note that, of the 106 nuclear plants operating during the period 1997 to 2001, only 27 (25%) experienced fuse failures. Figure 17 shows the distribution of age-related failures reported to the EPIX system per unit. Two units reported three fuse failures, six units reported two failures, and 19 units reported a single failure.

The failure modes identified for the fuse failures are shown in Figure 18. The dominant failure mode is "FUSE OPENS SPURIOUSLY" (73%), which is consistent with the results from previously discussed operating experience. Other failure modes include "INTERMITTENT/ ERRATIC OUTPUT" (16%); "FAIL TO OPERATE AS REQUIRED" (5%), and "HIGH RESISTANCE/LOSS OF CONTINUITY" (3%).

The predominant fuse failure mechanism was "FATIGUE/DEGRADATION OF THE FUSIBLE ELEMENT," which accounted for 62% of the failures. This is consistent with the predominant failure mechanism found in the NPRDS data. Again, for failures attributed to aging in which no information on the failure mechanism was presented in the report, and the fuse is described as "blown," it was assumed for purposes of this evaluation that the mechanism was fatigue of the fusible element. Other less frequently found mechanisms were "LOOSENING/DEGRADATION OF THE WIRING CONNECTIONS" (16%), "MECHANICAL WEAR/FATIGUE OF FUSE HOLDER CLIPS" which caused loosening of the fuse in the fuse holder (5%), and "LOOSENING/ DEGRADATION OF THE FUSE END CAPS" (5%). These results are summarized in Figure 19.

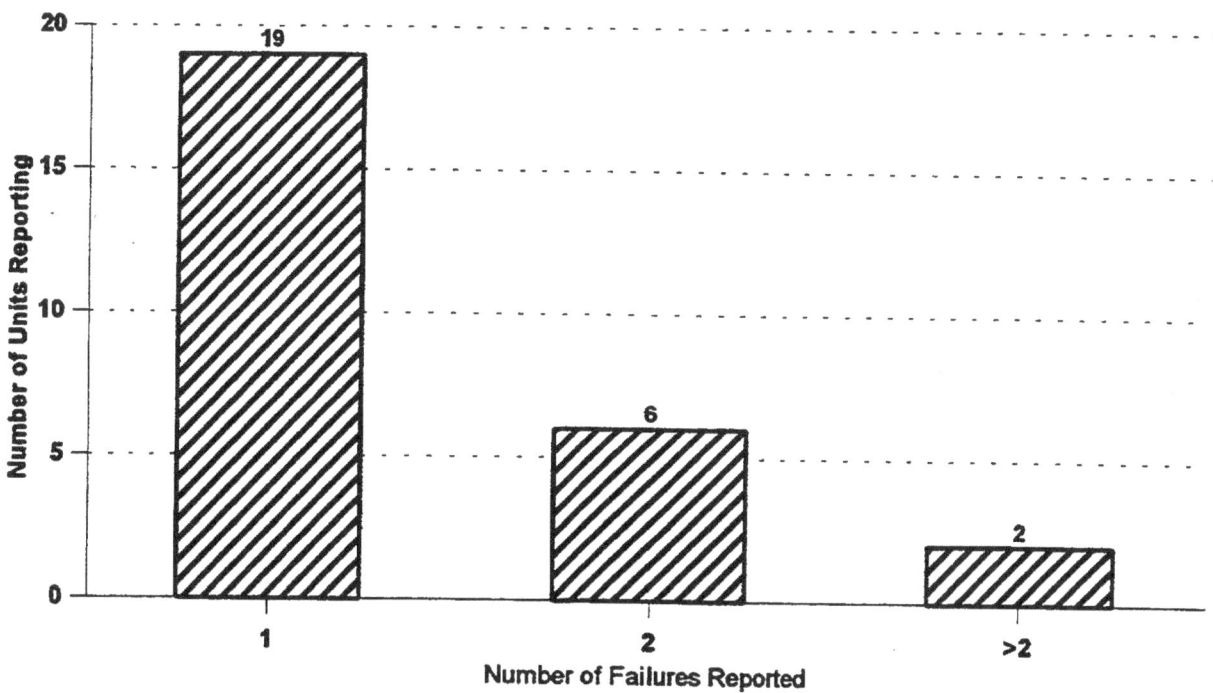

Figure 17 Distribution of age-related fuse failures per unit from EPIX

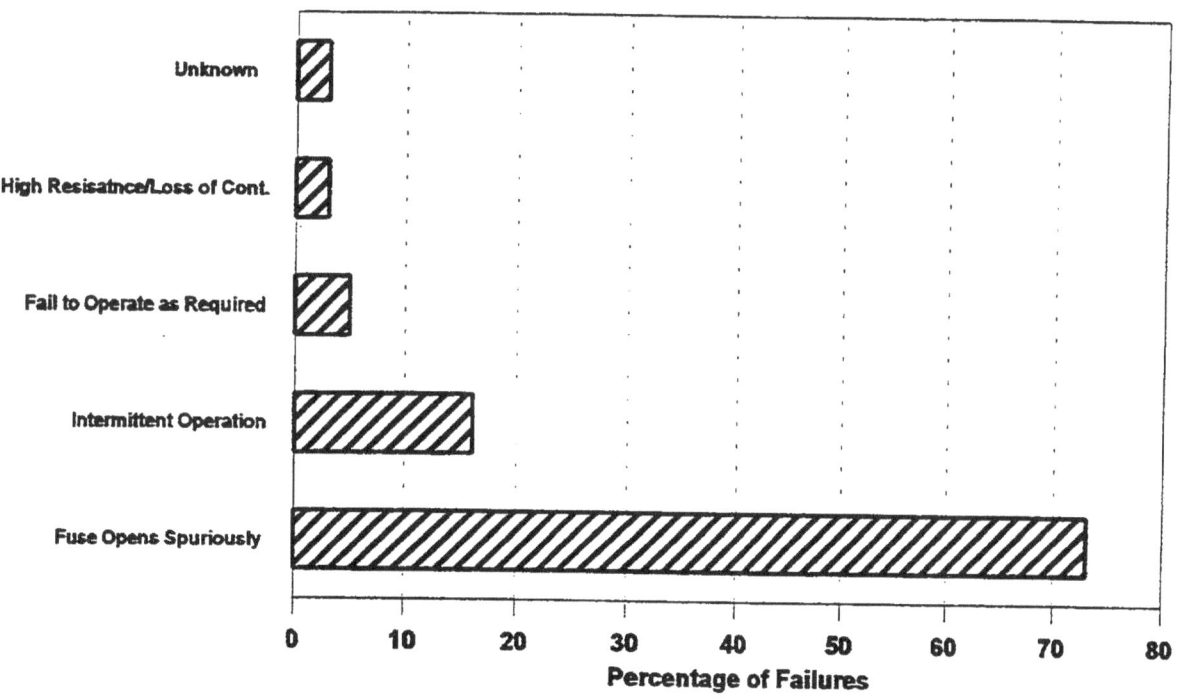

Figure 18 Distribution of fuse failure modes from EPIX

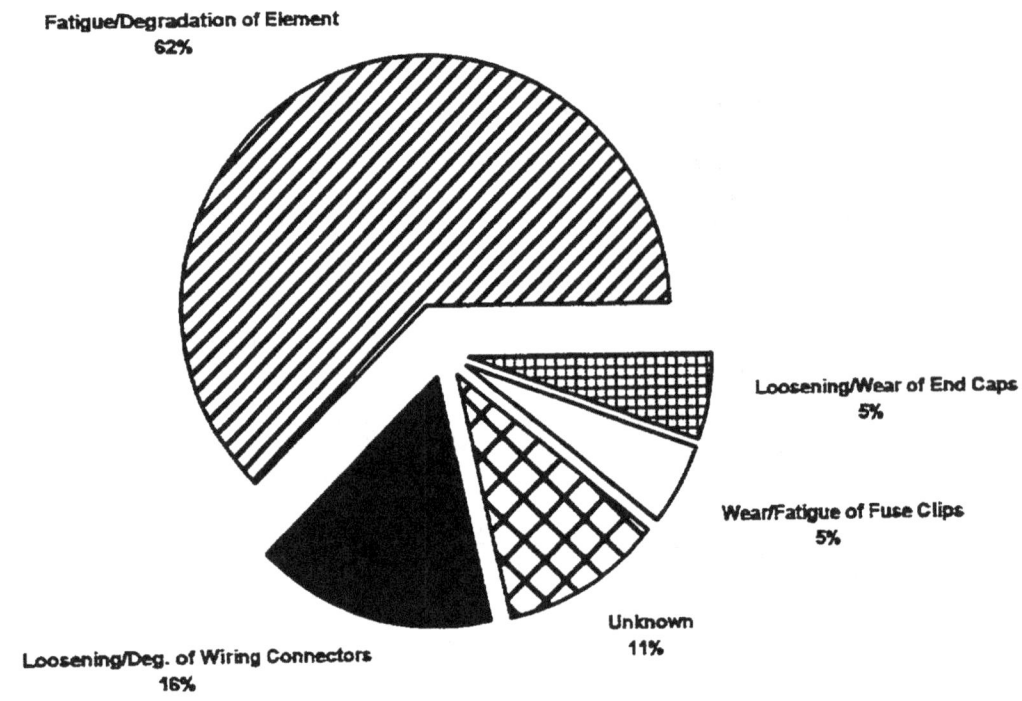

Figure 19 Fuse failure mechanisms from EPIX

It should be noted that between one quarter and one third of the EPIX and NPRDS failures involved mechanisms affecting the fuse holder assembly. The quantity found in the LER data was less than this, however, it was still significant at approximately 20%. The lower amount of fuse holder-related mechanisms among the LER failures can probably be attributed to the more frequent inspection and maintenance received by the safety-related systems that are represented in greater proportion in the LER data.

The effect of fuse failure on plant performance was not usually given in the EPIX system. Many of the fuse failures were detected during plant shutdown and the effect on the plant is not stated. For failures during plant operation, any reductions in power output or challenges to safety systems were seldom given. However, three fuse failures did cause plant shutdown. These were:

- EPIX Event at Nine Mile Point 2 on 11/11/98: A control rod dropped causing a plant scram during an operational safety procedure test on the reactor protection system. The rod drop was due to failure in a logic control circuit fuse. It was found during troubleshooting that voltage was present at one end of the fuse but not the other. A new fuse was installed and the problem was corrected. However, the old fuse, when tested, proved to be "good." It was concluded that the fuse holder was faulty, but this could not be confirmed by additional testing. The fuse holder was replaced.

- EPIX Event at Oconee Unit 2 on 02/28/99: During full-power operation, a small reactor transient occurred when power to the turbine control valve electro-hydraulic control system was lost and backup power took over. The problem was traced to a failure in a circuit breaker fuse that was connected to the turbine control valve circuit. While

38

investigating the fuse failure, spurious power was found to be conducting through the "failed" fuse. The reaction of the electro-hydraulic control system caused the turbine control valve to close. In turn, this caused the reactor to trip on high reactor coolant system pressure. The fuse was found to have failed because one end of the fusible link had become detached. Overcurrent failure had not occurred.

- EPIX Event at the South Texas 1 Plant on 5/16/99: A fuse on the primary side of a power transformer on a 13.8 kV bus began to degrade without the operators being aware. When the voltage drop on the secondary side reached the lower setpoint, it tripped a relay that controls the large electric motors that operate a circulating water pump and a reactor coolant pump in loop 3. The reactor tripped due to a low flow condition in the loop. The fuse failure was determined to be caused by aging/wearout.

In analyzing the subcomponent that failed in the fuse assembly, the "FUSIBLE LINK" is the major contributor (68% of the total failures). This is followed by the "FUSE HOLDER CLIPS" (22%), which failed as a result of fatigue, wear, and embrittlement. "CONNECTORS" failed 5% of the time and involved disconnection of the connector to the fuse holder and, in one case, fusion of the connector itself. Another 5% of the failures were attributed to the whole fuse assembly, which was deemed to be a generally faulty component. These results are summarized in Figure 20.

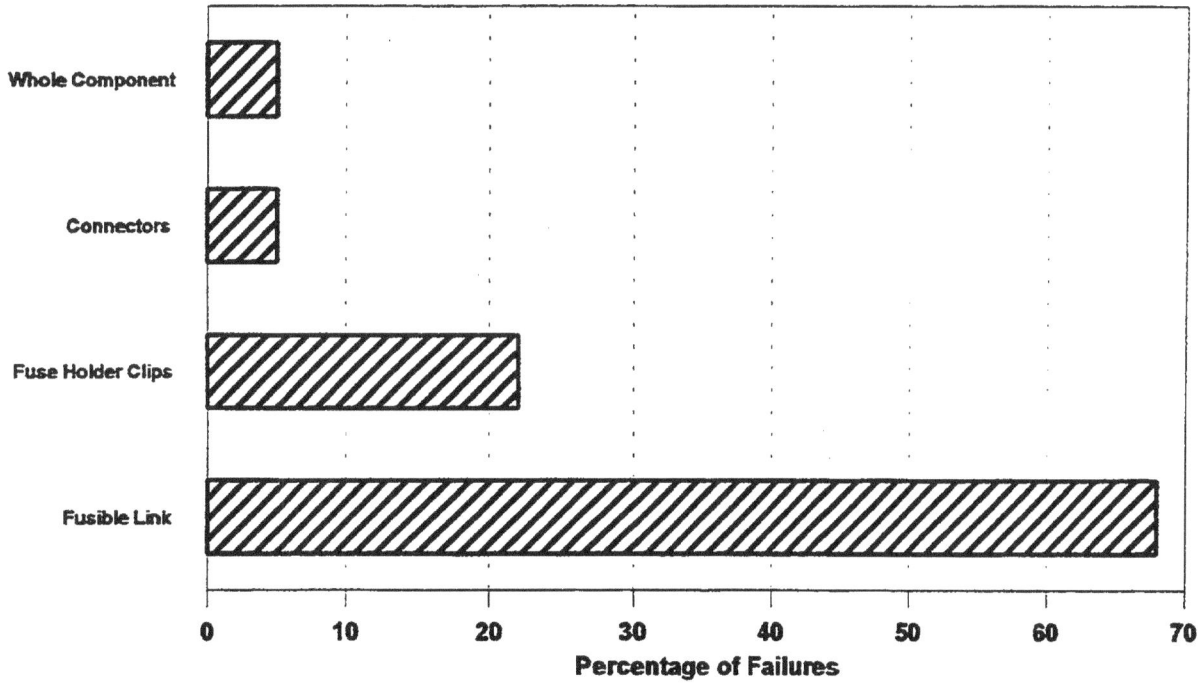

Figure 20 Distribution of fuse subcomponents failed from EPIX

The applications for which fuses failed by age-related mechanisms were analyzed in detail. In 22% of the EPIX listings, fuse failure resulted in the non-operability of valves. These fuses

39

were either in a logic controller or in a power circuit for the valve. Another 22% of the fuse failures caused failure of instrumentation circuits. The applications for the fuse failures are summarized in Figure 21.

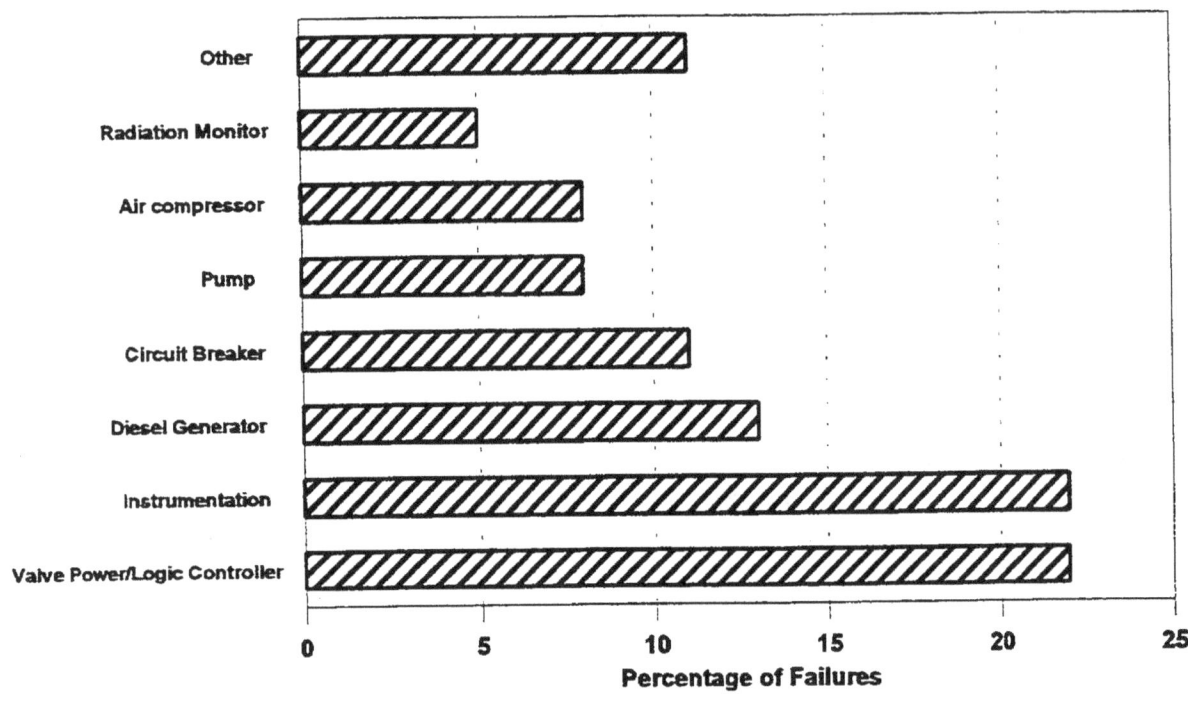

Figure 21 Distribution of applications for failed fuses from EPIX

Figure 22 shows the fuse failure detection methods that were identified from the EPIX system data. "OPERATIONAL ABNORMALITIES" detected 46% of the fuse failures, and a large fraction of failures were detected by "SURVEILLANCE TESTING" (27%). "ROUTINE OBSERVATION" (16%) and "AUDIOVISUAL ALARMS" (11%) were the remaining detection methods.

5.4   Trend Analysis of Operating Experience
To identify any trends in the failure rate of fuses, the data from each of the databases were analyzed to identify trends that could indicate potential aging concerns. The trend analyses are discussed in the following paragraphs.

5.4.1   LER Data Trend Analysis
The LER data were sorted to obtain fuse failures per year over the past ten years. If the number of fuse failures per year was increasing, this could indicate a potential problem with management of fuse aging. However, the results presented in Figure 23 show no discernible trend in the number of reported fuse failures. The relatively large number of fuse failures reported in 1991 is followed by a decline in 1992. This could reflect the increased attention to fuse control programs suggested in NRC Information Notice 91-51. At that time, a number of

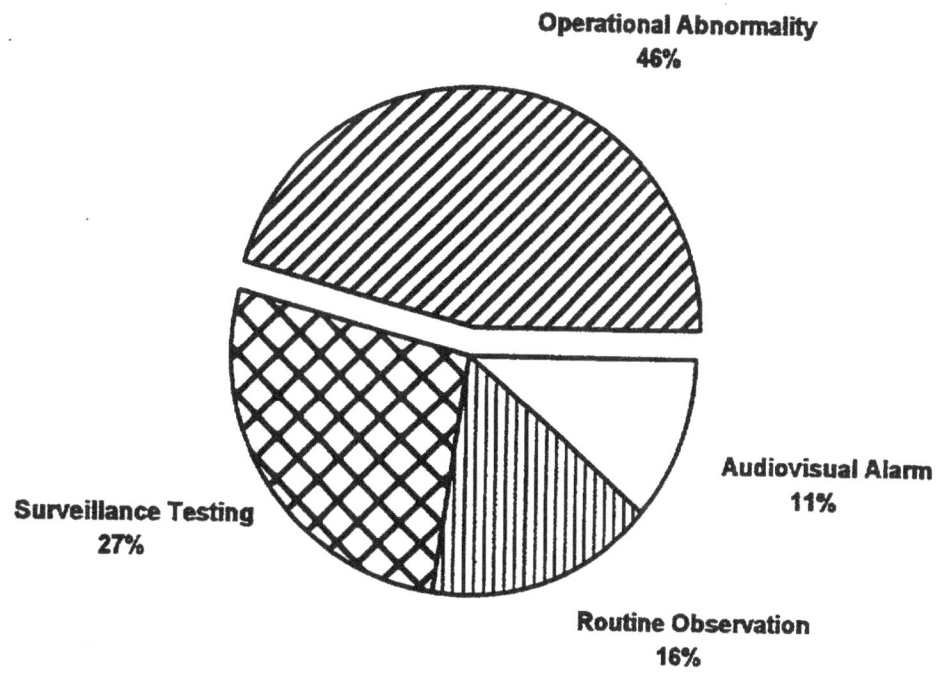

Figure 22 Fuse failure detection methods from EPIX

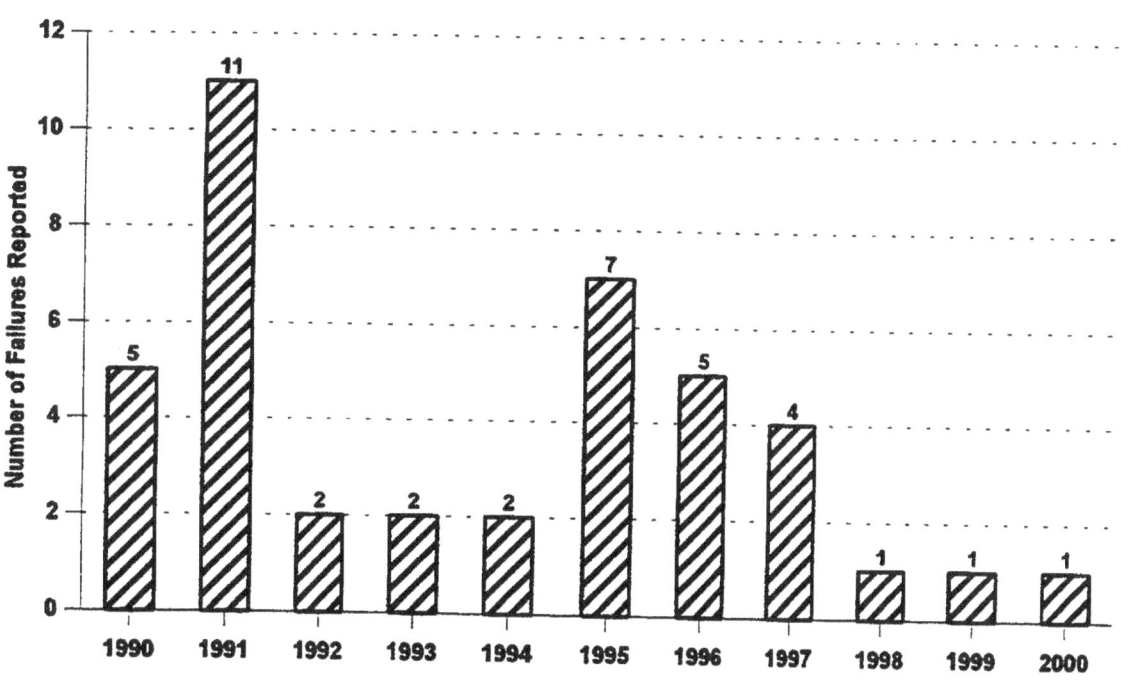

Figure 23 Number of age-related fuse failures reported per year as LERs

deficiencies were noted in fuse control programs that could have led to fuse failures. The notice concluded that a well designed fuse control program could decrease fuse problems significantly. This increased attention to fuse control could have led to the decline in fuse failures in 1992.

The number of fuse failures in 1995 and 1996 are also somewhat higher than other years. This could be related to the problems with fuse ferrule cracking and fuse element cracking noted in the Part 21 reports discussed previously. These generic fuse problems were identified in 1995 and 1996, respectively. After addressing these problems the number of fuse failures again declined. From 1997 to the present, an average of one fuse failure per year was reported as an LER, suggesting that age-related fuse failures are currently being controlled. From the review of LERs it was noted that whenever a fuse failure occurred, other similar fuses in the plant were inspected, and many times replaced as a precautionary measure. This could be one reason that fuse failures do not show an increasing trend.

5.4.2 NPRDS Data Trend Analysis

The data from the NPRDS regarding aging-related fuse failures were analyzed for any predominant trends. As mentioned previously, 1,145 of these were reviewed to provide the data for this study. The greater amount of fuse failure data available from the NPRDS database allowed for more detailed analysis for possible trends.

The aging-related fuse failures by year, as reported in the NPRDS database, are given in Figure 24, grouped by reactor type. The number of failures prior to 1984 is much lower than after that year for several reasons: 1) licensee reporting to the NPRDS was not comprehensive until 1984, 2) there were fewer operating nuclear units at that time, and 3) the reports selected for review in this study tried to emphasize the more recent operating history. The data show an increase in the number of age-related failures in 1990 and 1991 for both PWRs and BWRs similar to that observed for the LER data in Figure 23. As mentioned earlier, the heightened attention given to fuse inspections, maintenance, and control of replacements brought about by NRC Information Notice (IN) 91-51, probably contributed to the greater number of failures and other fuse problems discovered and reported during this time. Once programs and procedures for fuse inspection and surveillance were set in place by the licensees, they were in a better position to discover degraded fuse and/or fuse holder problems.

Figure 24 also shows a large number of fuse failures reported at PWRs in 1992 and 1993, and generally, after that time, a greater number of fuse failures than for BWRs. Since the total quantity of fuse failures reported in any given year would depend on the number of nuclear units that were operating during that year, the data were normalized by dividing the annual failure totals by the number of operating nuclear units in each year. The resulting plot of fuse failures per unit year showed that the rate of age-related fuse failures was nearly the same for BWRs and PWRs throughout the period except for the years 1991 and 1992, where the rate was more than double for PWRs. To examine this more closely, the data for PWRs was sorted by nuclear steam supply system (NSSS) vendor, and normalized by dividing the annual totals by the number of operating nuclear units in each year. The results of this analysis are presented in Figure 25.

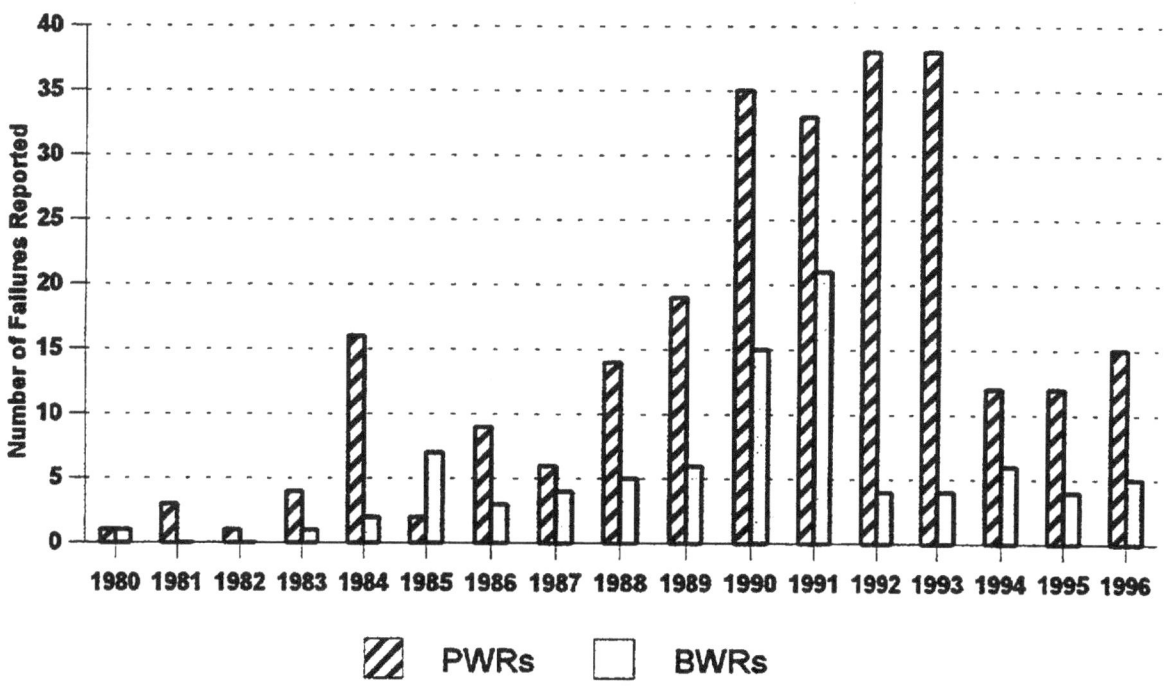

Figure 24 Number of age-related fuse failures reported per year to NPRDS

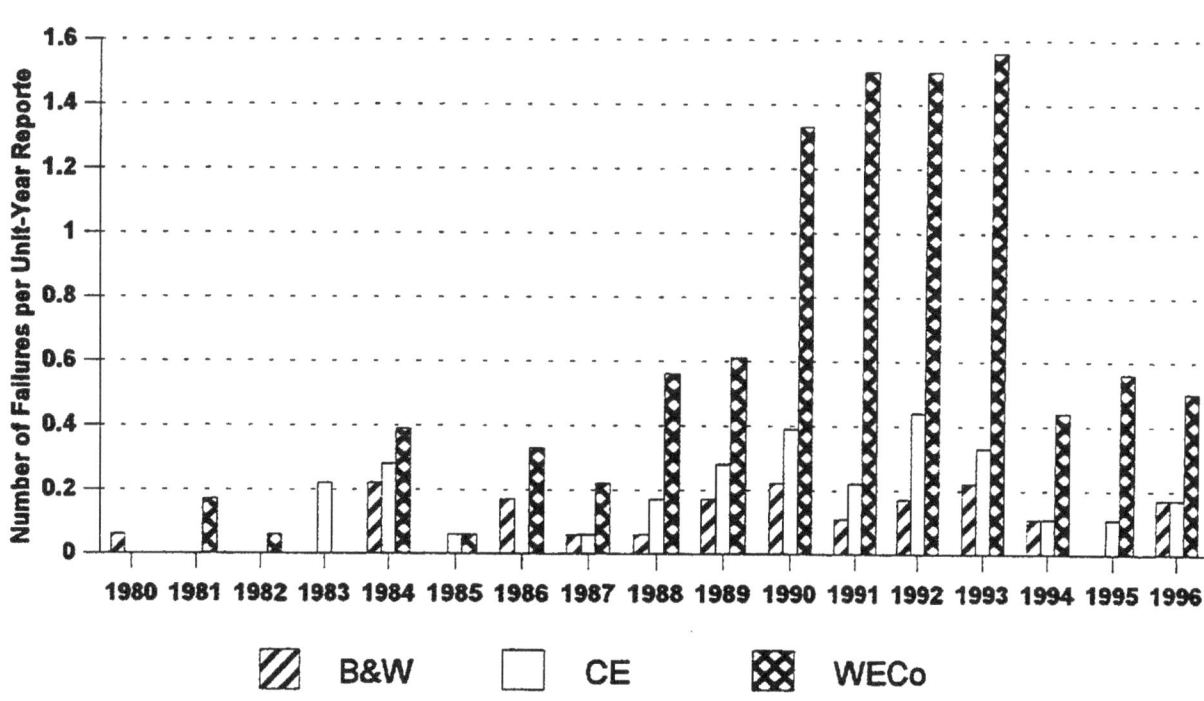

Figure 25 Number of age-related fuse failures per unit-year for PWRs reported to NPRDS

43

Figure 25 indicates that Westinghouse (WECo) PWRs had a higher rate of age-related fuse failures per unit year than the Combustion Engineering (CE) and Babcock & Wilcox (B&W) reactors throughout the period shown. During the years 1990 through 1993, the rate for Westinghouse reactors was more than four times that reported at the CE and B&W PWRs. Some of this difference may be attributed to the heightened awareness for fuse maintenance and inspection following IN 91-51 on fuse control programs. More likely, problems with incorrectly sized fuses in Westinghouse System 7300 printed circuit cards, as reported in NRC IN 93-87, contributed to the higher rate of age-related fuse failures per unit year. This is supported by the data which show that more than one third of the fuse failures reported in the years 1991-93 were in applications involving the process control system circuit cards or their power supply cards. From 1994 on, the number of failures per unit year for the Westinghouse plants dropped dramatically, however it still remained higher than for the CE and B&W units.

The NPRDS failure reports were next grouped by major fuse/fuse holder assembly subcomponents, and then by the age in service, in years, at the time each failure was discovered. The results of this analysis are displayed on Figure 26. As described previously in the review of operating experience, the largest number of fuse failures involved the fusible link, followed by fuse/fuse holder assembly failures. Figure 26 shows the fusible link failures to be particularly dominant from the about the fourth through the twelfth years. Fuse holder failures occurred at about the same number per year for approximately the first fifteen years. Similarly, the number of fuse holder wiring and connector failures is fairly steady throughout the first fifteen years of service. The latter is to be expected since wiring and connectors may be considered a part of the fuse holder assembly and are subject to the same aging mechanisms.

Note that the total number of age-related fuse failures reported in each year on Figure 26 decreases from left to right (increasing service age at time of failure discovery). This is not because older fuses are less likely to fail, but rather, that the population of fuses still in service at the longer lifetimes is becoming smaller and smaller. Insufficient data are available on the fuse populations in nuclear power plants and the service age of those populations to permit normalization of the raw failure quantities shown on Figure 26. Another potential problem with NPRDS data is that the "age at failure discovery" often refers to the age of the component or system in which a fuse is located. If fuses have been replaced one or more times over the life of the system, the "age at failure discovery" clock is not normally reset. Hence, the data at longer service life (greater than 12 to 15 years) will be less reliable for this type of analysis.

Another approach to searching for trends in the fuse failures data base was to look at all of the failures for a particular plant over an extended period of time. This was done for the Arkansas Nuclear One (ANO) plant for the period from 1980 through 1996 using the NPRDS database, and from 1997 to 2000 using the EPIX database. Since there are two nuclear units at this plant, a greater population of fuses could be examined in which the policies and procedures with regard to fuses would be similar. This review included nine age-related failures from ANO Unit 1 (out of 29 NPRDS reports reviewed), 15 age-related failures from ANO Unit 2 (out of 23 NPRDS reports reviewed), and one age-related failure from each unit from the EPIX data base. The age-related failures from the NPRDS and EPIX reviews were then grouped by the unit year of commercial operation (rather than by calendar year) in order to place the data from the two nuclear units on a comparable time scale. These resulting groupings are plotted on Figure 27.

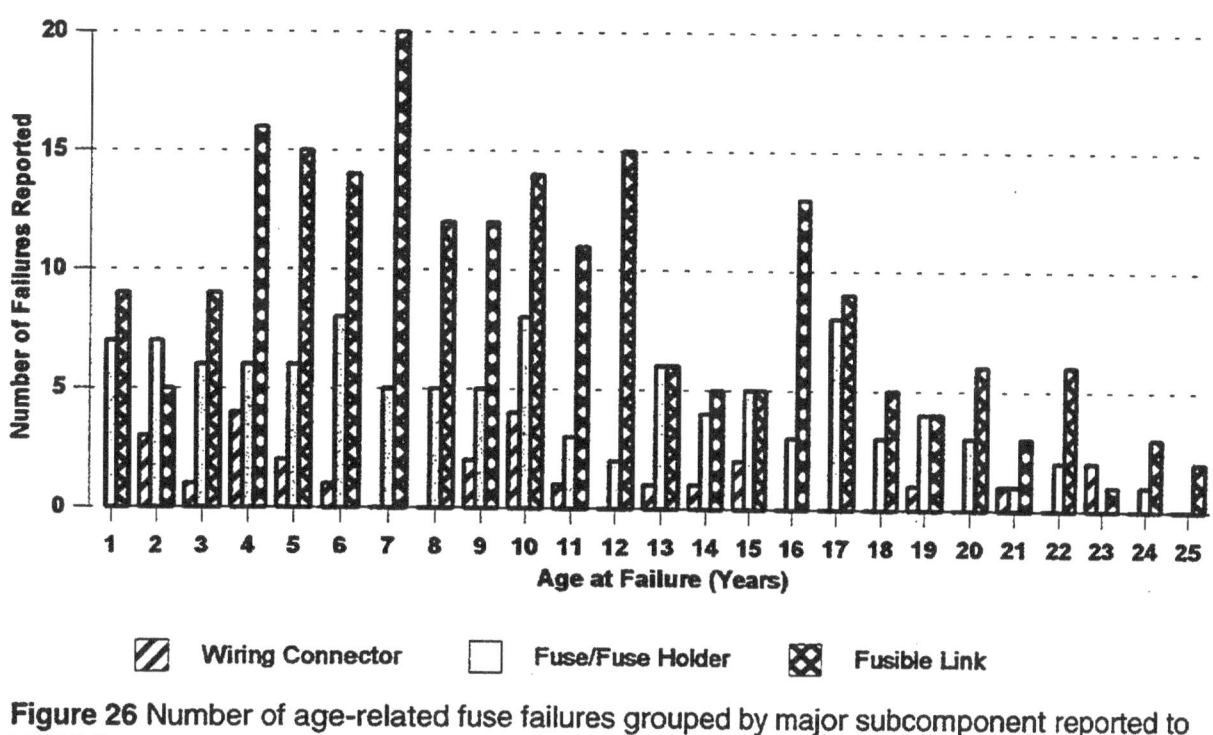

Figure 26 Number of age-related fuse failures grouped by major subcomponent reported to NPRDS

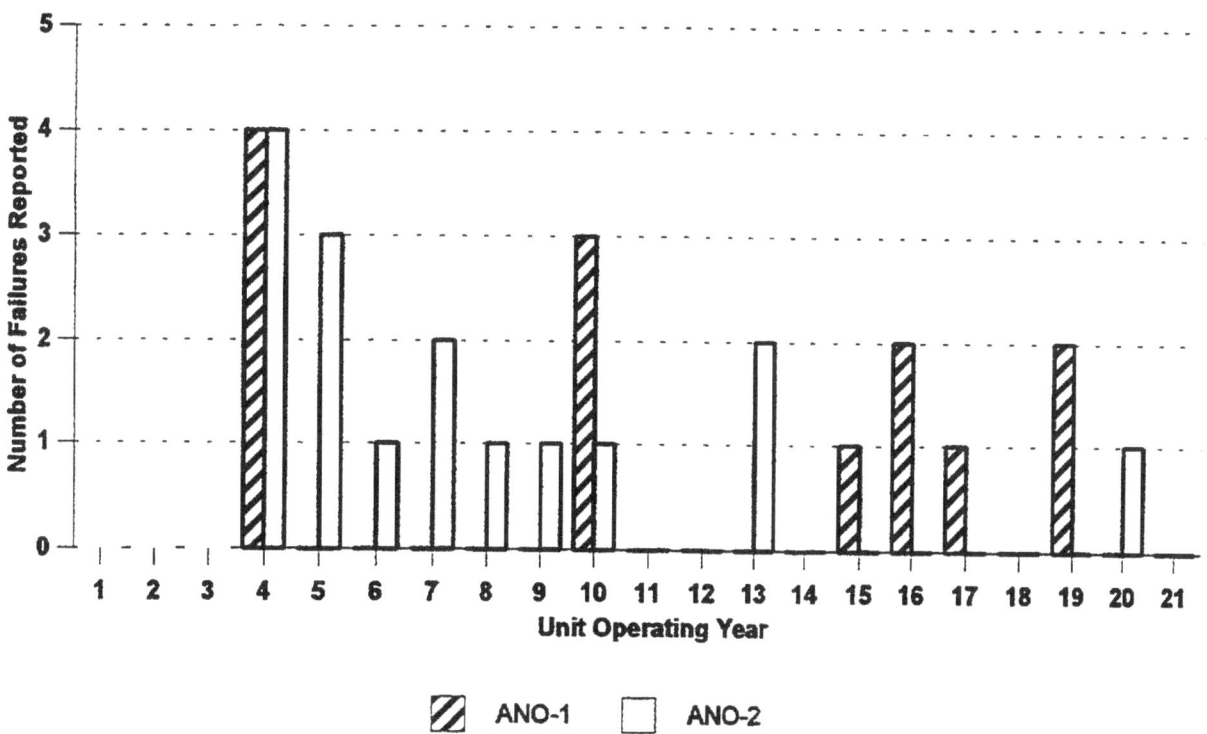

Figure 27 Age-related fuse failures at ANO 1 & 2 grouped by unit year - NPRDS & EPIX

From Figure 27 it can be seen that most of the failures for ANO Unit 2 occurred during the early years of the plant operation, during the fourth through the tenth years of operation. There were four age-related fuse failures at ANO Unit 2, occurring in four different systems, in the fourth year of operation. As seen in the figure, the number of failures at Unit 2 per unit year decreased over the next few years to the point where only sporadic fuse problems were reported after the tenth year of operation. One fuse failure in the fifth year of operation at Unit 2 occurred in a control power fuse to a control element assembly (reactor control rod drive) and resulted in a reactor trip.

In contrast, ANO Unit 1 experienced fewer failures overall than Unit 2, and, aside from four failures in the fourth year of operation, it reported no additional failures until the tenth year of operation. There were six fuse failures at Unit 1 between the fifteenth and nineteenth years of operation, and one in the twenty-third year of operation. One fuse failure in the sixteenth year of operation at Unit 1 occurred in an instrumentation and control power supply to a main feedwater system control valve that resulted in reduced power operation at Unit 1.

5.4.3 EPIX Data Trend Analysis

Fuse failures by year, as reported to the EPIX system are given in Figure 28. Since the EPIX system was initiated in 1997 to replace the NPRDS database, it contained only four years of operating experience. Therefore, extracting insights by trending this data is somewhat uncertain at the time this analysis was performed. Nevertheless, the data were evaluated for completeness. As shown, in the years 1997, 1998, and 1999, the number of age-related fuse failures was relatively constant, and no discernable trends are evident. In the year 2000, only two fuse failures occurred, which may be due to incomplete data entry into the database at the time of this study. The current year 2001 is not included in this evaluation since it is a partial year.

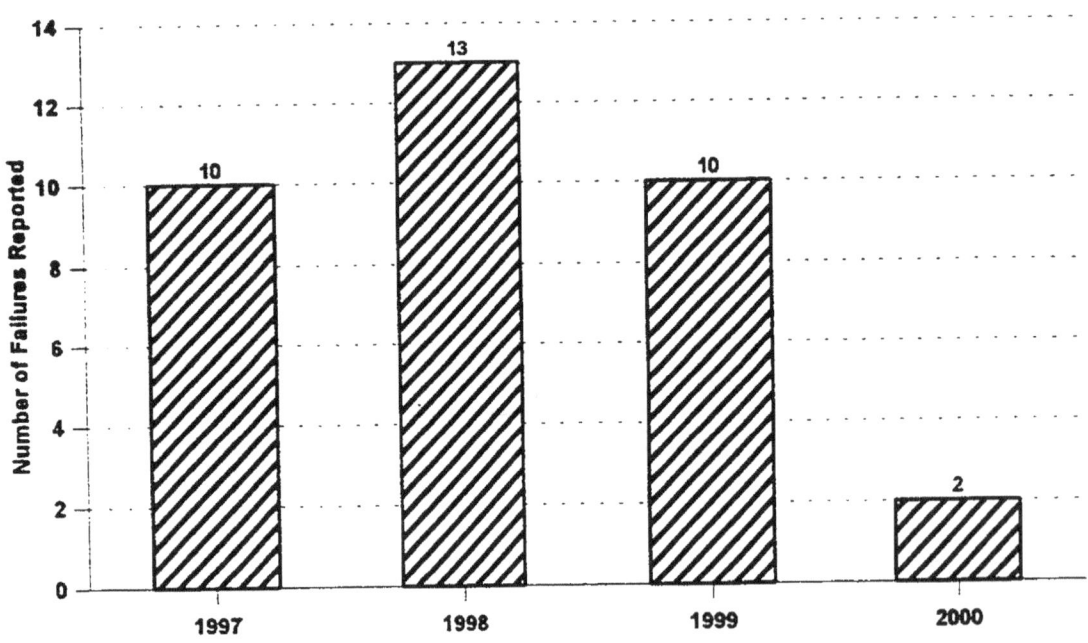

Figure 28 Number of age-related fuse failures reported per year to EPIX

46

6.0   Manufacturer's Insights

To supplement the information obtained from the review of past operating experience, the Cooper-Bussmann fuse manufacturing facility in Black Mountain, North Carolina was visited. The purpose of this visit was to obtain manufacturer's insights into the potential causes of aging degradation of fuses used in nuclear power plants, and to learn of any new developments in fuse designs and manufacturing that would mitigate the potential for aging degradation.

Discussions were held on the construction of fuses, and the potential types of aging degradation problems that could occur in service. Also, typical problems seen by the manufacturer from their customers were discussed. The following information was obtained:

- Materials of construction of the Bussmann fuses include the following
     - fuse tubes are typically made of glass (small diameter fuses), Melamine, paper, or a polyester material
     - fuse elements are typically zinc, copper, or silver
     - fuse ferrules and end caps for fuses rated for under 60 amps are 90% Copper/10% Zinc (this replaced the previously used 70Cu/30Zn used prior to 1980, which had problems with stress corrosion cracking). Fuses with higher current ratings use brass or stainless steel for the ferrules/end caps.
     - the larger fuses are packed with silica sand or a plaster of Paris material to minimize oxygen concentrations inside the fuse tube and provide structural support for the fuse element

- The fuse element designs vary based on the desired function of the fuse (i.e., one-time fuse, current limiting, or dual action fuse). The variations include the use of dual elements, changes in element cross-section, or application of tin lugs to the element. Fuse element cross sections are varied to achieve the desired opening characteristics of the fuse under short-circuit conditions. Tin lugs are applied to the elements to selectively weaken specific spots on the element during overload conditions (at temperatures of approximately 600°F, the tin will migrate into the copper and result in an alloy with a lower melting point).

- Testing of the fuses is limited to resistance testing and X-rays. The high precision resistance test is used during manufacture to determine if the fuse is acceptable. A specific resistance range is associated with an acceptable fuse in new condition. These resistance values can change with degradation of the fuse element and may be useful as an indicator of aging degradation in service. X-rays are used to verify fuse element integrity and to ensure that no excess solder is inadvertently deposited on the fuse element during manufacture, which would change the fuse performance properties. X-rays are also used in failure evaluations to provide information on the root cause of the fuse opening.

- Thermal imaging is a useful technique to locate "hot-spots," which are potential degradation sites. Typical hot-spots are loose connections and degraded contacts or connectors. If a hot-spot is detected, further investigation can be performed to determine if degradation of the fuse has occurred.
- Fuses are essentially thermal devices. As such, external elevated temperatures can

influence the performance of the fuse and change its rating. The higher the external temperature, the lower the fuse rating will be.

- Temperature cycling is a potentially significant aging stressor since the fuse element will expand and contract, and could be weakened due to work hardening. In some cases, with sand filled fuse tubes, the sand can shift during expansion/contraction of the fuse element and prevent the element from returning to its original position. This can impart mechanical stresses on the element and cause it to fail prematurely.

- Fuses that are operated continuously at less than approximately 60% of their rated current could potentially have an unlimited life. The worst case operating conditions would be in an application for which the fuse is repeatedly cycled from zero current to 90% or more of rated current. This would expose the fuse element to potentially severe mechanical stress due to expansion and contraction.

- Moisture intrusion is another potential aging stressor. Corrosion of the external fuse holder connections is possible, which could lead to higher resistance and heat buildup. Also, the fuses are not hermetically sealed and moisture intrusion can occur to the inner parts of the fuse. In locations with a high moisture and sulfur content, this could lead to corrosion of the internal metallic subcomponents of the fuse and result in changes in fuse performance or premature failure. This would be a concern for fuses in circuits that are not normally operating (e.g., long-term standby circuits) since there would be no heat generation to drive off moisture.

- Moisture is also a concern for fuses constructed with paper and fiber tubes. Under prolonged exposure to high humidity conditions, the paper tubes can absorb moisture causing them to experience dimensional changes. This phenomenon, if severe, could actually stretch the fuse element and cause it to break, leading to premature failure of the fuse. Once the fuse tube dries out, it may appear unchanged externally, however, it will seem that the fuse element had opened due to abnormal circuit conditions.

- Migration of zinc metal at elevated temperatures can be a concern for fuses that use zinc for the fuse element. At high temperatures, the zinc can actually migrate from one section of the fuse element to another. If the zinc migrates to a designed weak-link in the element, it can increase the cross section at that point and actually increase the rating of the fuse. Migration is not a problem with copper or silver fuse elements since these metals have been found to be dimensionally stable throughout the operating range of most fuses.

- Handling can be a concern for fuses with very fine elements. Bussmann manufactures a fuse for use in nuclear plant control rod drive mechanisms that has such an element. Special procedures must be implemented for the manufacture, testing, packaging, and handling of these fuses to safeguard their integrity.

Following the interview, a tour of the manufacturing plant was taken. The various stages of fuse manufacture were observed, along with the equipment and processes used to assemble

48

the fuses. The following insights were obtained from the tour:

- A number of very specialized robotic machines are used to assemble the fuses. Some machines assemble the entire fuse, test it and package it in one operation. Other machines assemble only parts of the fuse, after which it is transferred to another machine or a human operator for completion.

- Soldering of the fuse elements and end caps is performed in a variety of ways, depending of the type and size of fuse. The parts to be soldered can be heated electrically, with a gas flame, or in a salt bath (non-copper products). Non-acid fluxes are used to prevent corrosion of the soldered parts during service.

- Proper packing of the silica sand filler in the fuse tube is important for proper fuse performance, and great care is taken to ensure that the fuse is properly packed. Vibrating machines are typically used to ensure proper packing. Improper packing can result in voids or "columning" of the sand, which will degrade the fuse performance. In some specialty fuses, the fuses are heated in kilns to fuse the filler material into a solid.

- Attachment of the fuse end caps is performed in various ways depending on the fuse size. Small fuses have the cap held in place by wicking solder up into the cap. Larger fuses have the caps pressed on and, in some cases, the caps are notched and crimped to the fuse tube. Very large fuses can also have pins inserted to hold the caps in place.

New fuse designs were also discussed. Currently, a "box" fuse design is being produced by Bussmann. This design has all of the traditional fuse components enclosed in a plastic box so that it is "touch safe" (i.e., it has no exposed parts that are electrically energized). The box has prongs that are used to plug the fuse into a specially matched fuse holder. The materials of construction of the various fuse elements are the same as in traditional fuses, and this fuse design would be susceptible to similar aging mechanisms.

7.0   Summary and Conclusions

7.1   Summary

The operating experience reviewed herein provided a number of insights into the aging characteristics of fuses used in commercial nuclear power plant applications. Specific observations made from this review are summarized in the following:

Fuse Performance

*   Considering the thousands of fuses of all sizes and types that are in service at each of the 114 nuclear power plants that operated during the period of this study, the number of age-related fuse failures reported to the NPRDS, LER, and EPIX databases was relatively low. This indicates that an age-related fuse failure is an infrequent occurrence.

*   The operating experience data show that fuse failures can often go undetected until the system or component is called upon to operate. However, the designed-in redundancy of the affected instruments, components, and systems allows such failures to be tolerated with little or no effect on system or plant operation. Once they are identified, fuse failures normally are corrected easily and rapidly; NPRDS data show that more than half were corrected either on the same day or by the next day after they were discovered. In addition, the data indicate that, when an age-related fuse failure is identified, the utilities typically replace the fuses in all redundant trains for that application.

*   Fuses that are operated continuously at less than approximately 60% of their rated current could potentially have an unlimited life. The worst case operating conditions would be in an application for which the fuse is repeatedly cycled from zero current to 90% or more of rated current. This would expose the fuse element to potentially severe mechanical stress due to expansion and contraction.

*   The data were evaluated for trends that would indicate the degree to which aging degradation of fuses is being managed. The results show no discernible trend in the number of reported fuse failures. From 1997 to the present, an average of one fuse failure per year was reported as an LER, suggesting that age-related fuse failures are currently being controlled.

Fuse Aging Characteristics

*   The predominant failure mode experienced is "FUSE OPENS SPURIOUSLY," which indicates that the fusible element opened and caused an open circuit when it should not have. The failure mode "HIGH RESISTANCE/LOSS OF CONTINUITY" is also significant and represents a failure of the fuse holder. A small number of events involve "INTERMITTENT OPERATION" of the fuse, which are typically caused by loose fuse holder clips or faulty fuse holder wiring terminations. Finally, less than 1% of the events involve a "GROUND FAULT," which is usually caused by dirt or contamination on a high voltage fuse/fuse holder assembly faulting to ground.

- The predominant failure mechanism for fuses is "FATIGUE/DEGRADATION OF ELEMENT," i.e., of the fusible element or link, which leads to unexpected failure of the fuse. This is to be distinguished from the normal opening of the fusible link when the fuse is exposed to overcurrent conditions for a prescribed time. Fatigue is typically due to the degradation of the metallic fuse element over time as a result of exposure to elevated temperature, voltage transients, or short duration overcurrent conditions. It can lead to weakening of the fuse element, or a reduction in cross section, which reduces its current carrying capacity.

- A second important failure mechanism is "WEAR/FATIGUE OF FUSE CLIPS." Fatigue of the fuse holder clips can typically occur due to high temperature, mechanical stress, and repeated insertion and removal of the fuses, such as during maintenance or surveillance testing. Other less frequently cited failure mechanisms are "CORROSION/DEGRADATION OF CONTACT SURFACES," "LOOSE, BROKEN, OR DEGRADED WIRING CONNECTIONS," and "LOOSENING/WEAR OF END CAPS."

- Most fuse failures result in either "DEGRADED TRAIN/CHANNEL" or the "LOSS OF ONE OR MORE TRAINS/CHANNEL FUNCTIONS." Less frequent effects include "DEGRADED SYSTEM OPERATION," and the "LOSS OF ONE OR MORE SYSTEM FUNCTIONS." In 16% of the fuse failures reported to NPRDS, the plants indicated that there was "NO EFFECT" at all on the system in which the problem was found.

- The most common applications in which fuses fail are control power fuses for motor operated valves and dampers or solenoid operated valves. Also significant were electronic circuit card fuses for process control circuits and systems, and closely related were instrument power supply fuses. Finally, a fairly large number of fuse failures occurred in large power supply/rectifier applications, and in battery chargers/inverters/uninterruptible power supplies.

- The greatest number of fuse failures are detected during "MAINTENANCE" and "TESTING" activities, followed by "OBSERVATION" by plant operating and maintenance personnel. This is significant because it reflects proactive efforts on the part of licensees to find these failures before they can cause more serious problems. Other fuse failures are detected by "OPERATIONAL ABNORMALITIES." This category would include failures to operate when required, off-normal performance, loss of position indication, or loss of control power indication. Finally, "AUDIOVISUAL ALARMS" accounted for the detection of the remaining fuse failures. It should be noted that several reports mentioned that imaging infrared thermography surveys were used to identify incipient fuse and fuse holder failures. Licensees are taking advantage of this powerful inspection technology to detect the tell-tale hot spots that could indicate potential fuse assembly failures.

- Fuses are essentially thermal devices. As such, external elevated temperatures can influence the performance of the fuse and change its rating. The higher the external temperature, the lower the fuse rating will be.

- Temperature cycling is a potentially significant aging stressor since the fuse element will expand and contract, and could be weakened due to work hardening. In some cases, with sand filled fuse tubes, the sand can shift during expansion/contraction of the fuse element and prevent the element from returning to its original position. This can impart mechanical stresses on the element and cause it to fail prematurely.

- Moisture intrusion is another potential aging stressor. Corrosion of the external fuse holder connections is possible, which could lead to higher resistance and heat buildup. Also, the fuses are not hermetically sealed and moisture intrusion can occur to the inner parts of the fuse. In locations with a high moisture and sulfur content, this could lead to corrosion of the internal metallic subcomponents of the fuse and result in changes in fuse performance or premature failure. This would be a concern for fuses in circuits that are not normally operating (e.g., long-term standby circuits) since there would be no continuous heat generation to drive off moisture.

- Moisture is also a concern for fuses constructed with paper and fiber tubes. Under prolonged exposure to high humidity conditions, the paper tubes can absorb moisture causing them to experience dimensional changes. This phenomenon, if severe, could actually stretch the fuse element and cause it to break, leading to premature failure of the fuse. Once the fuse tube dries out, it may appear unchanged externally, however, it will seem that the fuse element had opened due to abnormal circuit conditions.

- Migration of zinc metal at elevated temperatures can be a concern for fuses that use zinc for the fuse element. At high temperatures, the zinc can actually migrate from one section of the fuse element to another. If the zinc migrates to a designed weak-link in the element, it can increase the cross section at that point and actually increase the rating of the fuse. Migration is not a problem with copper or silver fuse elements since these metals have been found to be dimensionally stable throughout the operating range of most fuses.

Fuse Monitoring

- Testing of the fuses is limited to resistance testing and X-rays. The resistance test is used during manufacture to determine if the fuse is acceptable. A specific resistance value is associated with an acceptable fuse in new condition. These resistance values can change with degradation of the fuse element and may be useful as an indicator of aging degradation in service. X-rays are used to verify fuse element integrity and to ensure that no excess solder is inadvertently deposited on the fuse element during manufacture, which would change the fuse performance properties. X-rays are also used in failure evaluations to provide information on the root cause of the fuse opening.

- Thermal imaging is a useful technique to locate "hot-spots," which are potential degradation sites. Typical hot-spots are loose connections and degraded contacts or connectors. If a hot-spot is detected, further investigation can be performed to determine if degradation of the fuse has occurred.

7.2   Conclusions

Based on the information reviewed, the following conclusions are drawn regarding aging of fuses:

- This study has found that fuses are susceptible to aging degradation that can lead to failure, however, the occurrence is infrequent. In several cases the failures have had significant impact on plant performance, including loss of redundant safety systems, challenges to engineered safety features, or reactor trips. Fuse failures in non-safety systems, such as, main feedwater, condensate, and service water, can also have an impact on plant performance. However, complete loss of equipment or system function does not normally occur due to age degradation of fuses.

- The data indicate that the incidence of fuse failures is not increasing with age presently, indicating fuse aging is being managed.

- Currently, there are several methods available for monitoring fuses, namely, high precision resistance testing, which is used during manufacture to determine acceptability of the fuse, in situ visual inspection, and thermography, which is used in the field to locate hot-spots (potential degradation sites).

- Field inspections should include examination of fuse holders since these components account for a significant number of fuse failures due to loosening of the holder clips or electrical connections. Maintenance procedures should be reviewed to minimize the removal and reinsertion of fuses to de-energize components since this can lead to degradation of the fuse holders. Fuses that must be removed and inserted frequently for maintenance and surveillance should be included in periodic maintenance and inspection programs to monitor and control the effects of these repetitive activities on the fuse and fuse holder.

- Fuses with fragile elements should be identified in the field to assure they are handled properly during maintenance or repair activities to prevent inadvertent damage to the element.

- Fuses used in applications in which they are exposed to repeated cycles from zero load to full load should be monitored since they are susceptible to premature degradation and potential early failure due to repeated expansion and contraction of the fuse element.

- Fuses constructed with paper cartridges and used in humid environments should be monitored since they are susceptible to premature degradation and potential early failure due to moisture intrusion or swelling.

8.0   References

1.   Kueck, J.D., and Brinkley, D.W., Improving Fuse Reliability in Critical Control Circuits," IEEE Transactions on Energy Conversion, Vol. 5, No.3, pp 603-606. The Institute of Electrical and Electronics Engineers, Inc., New York, September 1990.

2.   Beaujean, D.A., Newbery, P.G., and Jayne, M.G., "Long-Time Operation of High Breaking Capacity Fuses," IEE Proceedings-A, Volume 140, No. 4., The Institution of Electrical Engineers. London, United Kindom, July 1993.

Appendix A: Fuse Types, Categories and Classification

A.1   Generic Fuse Characteristics

The term *fuse* is defined in ANSI/IEEE Std. 100-1984 [1] as "an overcurrent protective device with a circuit-opening fusible part that is heated and severed by the passage of overcurrent through it." The definition indicates that a fuse is responsive to circuit current levels and will provide a system with overcurrent protection [2]. The fusible element opens in a time that varies approximately with the square of the magnitude of the current that passes through it. The time-current characteristic of the fuse depends on its type and rating [3].

The most important aspects of fuses are the following [4, 5, 6]:

- *Ampere Rating* - the current that a fuse will carry continuously without significant immediate deterioration and without exceeding its temperature rise limits,

- *Voltage Rating (ac and/or dc)* - the maximum open circuit voltage in which a fuse can be used, yet safely interrupt an overcurrent,

- *Interrupting Rating* - the highest current at rated voltage that an overcurrent protective device is intended to interrupt under standard test conditions,

- *Physical Dimensions* - size of the cartridge,

- *Current Limiting Factors* - maximum instantaneous peak let-through current (I_P) that passes through a fuse during the total clearing time; let-through energy (I^2t) is a measure of the heat energy developed within a circuit during the fuse's clearing time,

- *Renewable or Nonrenewable Construction* - for Class H fuses only, these fuses are designed with a single renewable element that may be replaced after a fuse has opened, allowing the fuse cartridge assembly to be reused, and

- *Labeling* - identification on the fuse of requirements for current limitation

Non-time delay fuses are fuses that have no intentional time delay designed into their operating characteristic. They are generally employed in applications other than motor circuits or in combination with a circuit breaker in which the circuit breaker provides protection for overload currents and the fuse provides short circuit current protection [3].

Time delay fuses have intentional built-in time delays in the overload current range to allow temporary and harmless inrush currents to pass without opening the fuse, but are designed to open on sustained overloads and short circuits [6]. This characteristic allows the selection of a fuse whose current rating is closer to the full load current of the application. Time delay fuses are widely used in motor circuits because the time delay permits their use as running overcurrent protection [3].

A dual-element fuse is a special design that utilizes two individual elements in series inside the fuse tube. One element, such as a spring-actuated trigger assembly, will operate on overloads up to five to six times the fuse current rating. The other element, the short circuit section, operates on short circuit currents up to its interrupting rating [6]. A dual element time delay fuse thus provides protection for both motors and circuits, and makes it possible to use a fuse whose current rating is close to the full load current of the circuit. The fuse will permit the starting inrush current of a motor to pass, but will open to protect the circuit if a continuous overcurrent condition exists.

A.2   Low Voltage Fuses

Low- voltage fuses can generally be divided into four categories:

- *Cartridge fuses* which are designed for the protection of power, lighting and branch circuits in accordance with the National Electric Code. These fuses include fuses in the Underwriters Laboratories (UL) Classes H, G, K-1, K-5, K-9, J, L, RK1, RK5, T, and CC. Most of the low-voltage power, control, and branch circuit protective fuses found in nuclear power plants are of this category.

- *Special fuses* are designed for supplementary overcurrent protection where branch circuit or equivalent circuit protective applications are not involved. They are normally used for the protection of electrical equipment, such as capacitors, rectifiers, and integrally fused circuit breakers. These may be found in certain types of electrical equipment in nuclear power plants that use these components such as battery chargers, inverters, or uninterruptable power supplies.

- *Supplementary overcurrent protection fuses* are small fuses that are used where branch circuit or equivalent circuit protective applications are not involved. They are used mainly to protect electrical appliances and small electronic equipment. These may be found in nuclear power plant electrical systems and equipment.

- *Plug fuses* are screw-in fuses that are primarily used in older residential and small commercial load centers. These are not found in nuclear power plant applications and will not be included in this analysis.

The National Electrical Manufacturers Association (NEMA) and the Underwriters Laboratories (UL) have set standards that govern the manufacture, performance, size, and testing of fuses. Most fuse types are labeled and identified by their UL classification which are summarized in Figure A-1. The most common classes of fuses for nuclear power plant applications are discussed in the following paragraphs.

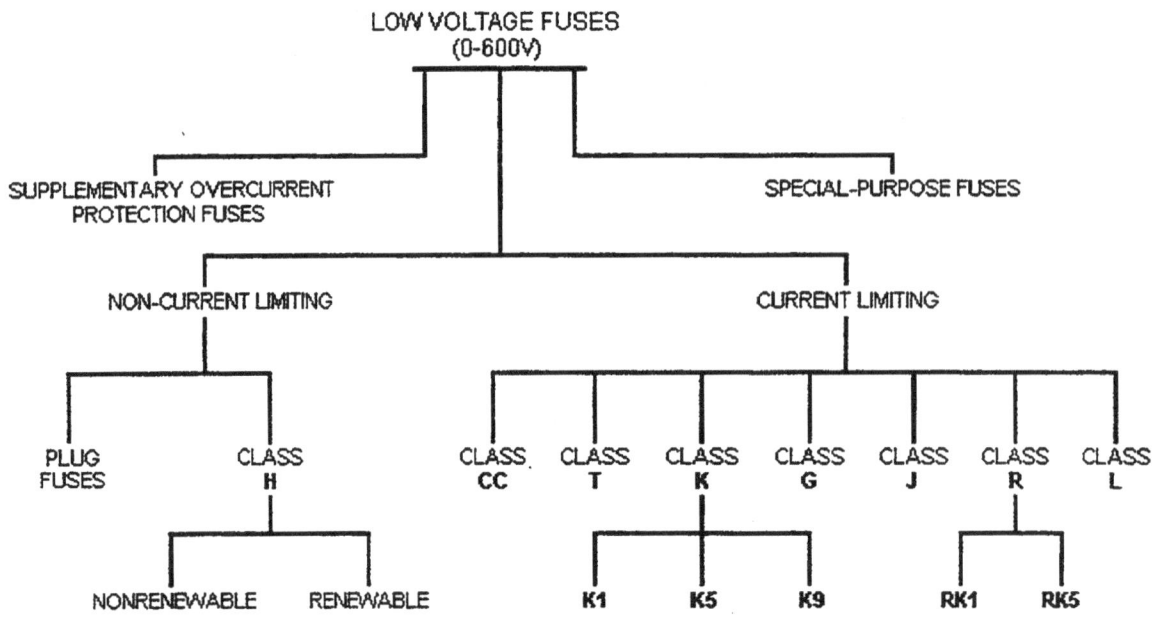

Figure A-1 Underwriters Laboratories, Inc. fuse classifications

Class H fuses - Cartridge fuses were formerly known as "NEC-dimensioned fuses." Class H fuses are tested and listed by Underwriters Laboratories in accordance with the UL standards [7, 8, 9] in 250V and 600V ratings with an interrupting capability of 10,000A. Class H fuses may be of the nonrenewable type, or of the renewable type in which the single element link may be replaced after the fuse is opened allowing the rest of the fuse assembly to be reused. The UL standards for renewable Class H fuses require a time delay of at least 10 seconds at five times rating in order carry the words "time delay" on the label. Renewable Class H fuses cannot be designated as time delay because the renewable link, as a single element, can't be designed to perform the melting function required to achieve time delay characteristics. Class H fuses are not marked as current limiting. Class H fuses are commonly used in generating station control circuits [10] and they are the most misapplied type of fuse in the electric industry [4, 11].

Class K fuses - These fuses meet UL standards [12, 13] for Class K in one of three sub-classes based on their maximum peak let-through current (I_p) and let-through energy (I^2t) (from lowest to highest): K1, K5, and K9. Dimensionally the same as Class H fuses, these fuses have no UL-recognized rejection feature. Their ac interrupting rating, up to 200kA, is marked on their labels, however, they are not permitted to be labeled as "current limiting" because of their interchangeability with Class H fuses, which are non-current limiting. They may be labeled as "time delay," "dual element," or simply "D" if the manufacturer has met the UL testing requirements for these optional features. The use of Class K fuses permits equipment and circuits to be applied on systems having potential fault currents in excess of the 10,000A of the dimensionally similar Class H fuse. Problems can arise from the dimensional similarity among fuses in the various Class K subclasses, Class R subclasses, and the lower interrupting rated Class H fuses [4, 11].

Class R fuses - These fuses meet UL standards [14, 15] for Class R in two sub-classes based on their maximum peak let-through current (I_P) and let-through energy (I^2t) (from lowest to highest): RK1 or RK5. These fuses have dimensions that provide a one-way physical rejection feature to prevent the substitution of fuses of any other class into equipment designed for Class R fuses. However, Class R fuses can be installed into older Class H or Class K equipment as a replacement to upgrade these systems to the maximum allowed by other devices in the system. Class R fuses are available with or without time delay and may be so marked in accordance with the UL standards [11].

Class J fuses - These fuses are rated to interrupt 200kA and meet UL standards [16, 17] for Class J fuses. They are UL labeled as "current limiting," are rated for 600Vac or less, and are of dimensions not interchangeable with any other class of fuses. Class J fuses that have a time delay of at least 10 seconds at five times rated current may carry a "time delay" marking on the label [11].

Class L fuses - These fuses have ratings in the range 601A-6,000A, meet UL standards [16, 17] for Class L fuses, are rated to interrupt 200kA ac, are rated for 600Vac or less, and have dimensions larger than those of other fuses rated 600V or less. They are intended to be bolted to electrical bus bars and are not used in clip-type holders. UL has no definition of time delay for Class L fuses, however many Class L fuses have substantial overload time-current carrying capability and sometimes are marked as "time delay" [4]. Class L fuse standards do not include 250V ratings, dc testing, nor dc ratings [11].

Class G fuses - Class G fuses, which meet the requirements of the applicable UL standards [16, 19] and are rated 300Vac to ground with interrupting ratings up to 100kA, were developed for use in lighting and appliance panelboards with a special fusible switch unit. These are nonrenewable cartridge fuses, with a maximum current rating of 60A, and dimensions that prevent interchangeability with any other class of fuses. Class G fuses are current limiting and are labeled as such per UL standards, which prescribe the peak let-through current (I_P) and let-through energy (I^2t). An optional time delay test included in the Class G fuse standard, calling for a minimum opening time of 12 seconds at 200% of the fuse's current rating, is unique to fuses in Classes G and CC [4].

Class T fuses - Class T fuses, which conform to UL standards [20, 21] for this class of fuses, are rated ac only, nonrenewable, current limiting, and are designed specifically for protection of feeders and branch circuits as prescribed in the NEC [22]. Current ratings range from 0 to 1200A for 300V rated fuses and 0 to 800A for 600V rated fuses. Their interrupting rating is 200kA symmetrical, with peak let-through current (I_P) and let-through energy (I^2t) specified for the individual case sizes. As mentioned for Class L fuses, Class T fuses also must be labeled "current limiting" and they may be labeled "time delay." The time delay characteristics of Class T fuses, are formally investigated by UL and must meet the optional time delay test requirements [4].

Class CC fuses - Class CC fuses are a nonrenewable, current limiting fuse intended for protection of components sensitive to short time overloads (such as, semiconductor applications and electronic equipment), noninductive loads (lighting and resistance heating), and short-circuit protection of motor circuits. UL standards [16, 23] for this class of fuses list current ratings up to 30A , rated 600Vac only, with an interrupting rating of 200kA symmetrical. Class CC fuses must be labeled "current limiting" and they may be labeled "time delay." Class CC fuses are current limiting and are labeled as such per the UL standards, which prescribe the peak let-through current (I_p) and let-through energy (I^2t). An optional time delay test included in the Class CC fuse standard, calling for a minimum opening time of 12 seconds at 200% of the fuse's current rating, is unique to fuses in Classes G and CC [4].

Supplementary overcurrent protection fuses - UL standards for supplementary overcurrent protection fuses [24, 25] cover three types of such fuses: micro fuses, miniature fuses, and miscellaneous fuses. The first two types are very small fuses used for protection of individual pieces of equipment or internal components and circuits of large packaged equipment. The miscellaneous fuses are usually ferrule type cartridge fuses that cannot be installed in fuseholders intended for Class H, K, R, J, L, G, and T. Maximum current rating is 30A, ac only, and voltage ratings range from 125Vac to 600Vac. The interrupting rating at 125V is 10,000A; optional interrupting ratings of 50kA and 100kA are available at the higher voltage ratings [4].

Special purpose fuses - This category is not covered by a specific UL standard. Special purpose fuses have operating characteristics that provide unique overcurrent and/or short circuit protection for such equipment as capacitors, rectifiers, and integrally fused circuit breakers. Current ratings range up to 6,000A and voltage ratings range up to 1,000V, placing some of these fuses into the medium voltage (greater than 600V) category. Interrupting ratings up to a maximum of 200kA symmetrical can be found with current limiting characteristics that are engineered to limit the let-through energy (I^2t) to levels that can be safely withstood by the specified protected equipment [4].

A.3   Medium Voltage Fuses

Power fuses found in nuclear power plants that are rated at over 600Vac are considered medium-voltage fuses and will be of either the current limiting type or the expulsion type. The characteristics and requirements for medium voltage power fuses can be found in ANSI C37.46-1981 [26]. As described in the standard, current limiting type fuses will be either general purpose (E-rated or non-E-rated) or R-rated. The general purpose fuse operates over a wider range of overcurrent levels as compared to the R-rated fuse, which is specifically intended to interrupt high magnitude fault currents.

The current limiting fuse is designed so that the melting of the fuse element introduces a high electric arc resistance into the circuit prior to the realization of the prospective maximum peak current of the first half-cycle of fault current. If the magnitude of the fault current is sufficiently high, the high resistance of the arc will effectively limit the peak value of fault current, i.e., the maximum instantaneous peak let-through current (I_p), that passes through the fuse into the protected circuit or equipment.

A.4   References

1.   ANSI/IEEE Std. 100-1984, "IEEE Standard Dictionary of Electrical and Electronics Terms," Institute of Electrical and Electronics Engineers, Inc. New York, 1984.

2.   IEEE Std 141-1986 (Red Book), "IEEE Recommended Practice for Electric Power Distribution for Industrial Plants," The Institute of Electrical and Electronics Engineers, Inc. New York, 1986.

3.   IEEE Std 241-1990 (Gray Book), "IEEE Recommended Practice for Electric Power Systems in Commercial Buildings," The Institute of Electrical and Electronics Engineers, Inc. New York, 1990.

4.   Reichenstein, Hermann W. and DeDad, John A., "Practical Guide to Applying Low-Voltage Fuses-Classes and Characteristics," EC&M Books, INTERTEC Electrical Group. Overland Park, Kansas, April 2001.

5.   Littelfuse Form No. 1FC-783-1, "Industrial Fuses," Littelfuse, Inc. Des Plaines, Illinois, undated.

6.   "SPD Electrical Protection Handbook-Selecting Protective Devices Based on the National Electric Code," Cooper Bussmann, Cooper Industries, Inc. St. Louis, Missouri, 2000.

7.   ANSI/UL 198B-1995, "Standard for Safety for Class H Fuses," Underwriters Laboratories, Inc. Northbrook, Illinois, 1995.

8.   ANSI/UL 248-6-2000, "Standard for Safety for Low-Voltage Fuses-Part 6: Class H Non-Renewable Fuses," Underwriters Laboratories, Inc. Northbrook, Illinois, 2000.

9.   ANSI/UL 248-7-2000, "Standard for Safety for Low-Voltage Fuses-Part 7: Class H Renewable Fuses," Underwriters Laboratories, Inc. Northbrook, Illinois, 2000.

10.   Kueck, J.D., and Brinkley, D.W., Improving Fuse Reliability in Critical Control Circuits," IEEE Transactions on Energy Conversion, Vol. 5, No.3, pp 603-606. The Institute of Electrical and Electronics Engineers, Inc., New York, September 1990.

11.   IEEE Std 242-1986 (Buff Book), "IEEE Recommended Practice for Protection and Coordination of Industrial and Commercial Power Systems," The Institute of Electrical and Electronics Engineers, Inc. New York, 1986.

12.   ANSI/UL 198D-1995, "Standard for Safety for Class K Fuses," Underwriters Laboratories, Inc. Northbrook, Illinois, 1995.

13.   ANSI/UL 248-9-2000, "Standard for Safety for Low-Voltage Fuses-Part 9: Class K Fuses," Underwriters Laboratories, Inc. Northbrook, Illinois, 2000.

14.   ANSI/UL 198E-1988, "Standard for Safety for Class R Fuses," Underwriters

Laboratories, Inc. Northbrook, Illinois, 1988.

15.   ANSI/UL 248-12-2000, "Standard for Safety for Low-Voltage Fuses-Part 12: Class R Fuses," Underwriters Laboratories, Inc. Northbrook, Illinois, 2000.

16.   ANSI/UL 198C-1986, "Standard for Safety for High Interrupting-Capacity Fuses, Current-Limiting Types" Underwriters Laboratories, Inc. Northbrook, Illinois, 1986.

17.   ANSI/UL 248-8-2000, "Standard for Safety for Low-Voltage Fuses-Part 8: Class J Fuses," Underwriters Laboratories, Inc. Northbrook, Illinois, 2000.

18.   ANSI/UL 248-10-2000, "Standard for Safety for Low-Voltage Fuses-Part 10: Class L Fuses," Underwriters Laboratories, Inc. Northbrook, Illinois, 2000.

19.   ANSI/UL 248-5-2000, "Standard for Safety for Low-Voltage Fuses-Part 5: Class G Fuses," Underwriters Laboratories, Inc. Northbrook, Illinois, 2000.

20.   ANSI/UL 198H-1988, "Standard for Safety for Class T Fuses," Underwriters Laboratories, Inc. Northbrook, Illinois, 1988.

21.   ANSI/UL 248-15-2000, "Standard for Safety for Low-Voltage Fuses-Part 15: Class T Fuses," Underwriters Laboratories, Inc. Northbrook, Illinois, 2000.

22.   ANSI/NFPA 70-1999, "National Electrical Code (NEC)," National Fire Protection Association. Quincy, Massachusetts, 1999.

23.   ANSI/UL 248-4-2000, "Standard for Safety for Low-Voltage Fuses-Part 4 : Class CC Fuses," Underwriters Laboratories, Inc. Northbrook, Illinois, 2000.

24.   ANSI/UL 198G-1988, "Standard for Safety for Fuses fur Supplementary Overcurrent Protection," Underwriters Laboratories, Inc. Northbrook, Illinois, 1988.

25.   ANSI/UL 248-14-2000, "Standard for Safety for Low-Voltage Fuses-Part 14: Supplemental Fuses," Underwriters Laboratories, Inc. Northbrook, Illinois, 2000.

26.   ANSI C37.46-1981, "American National Standard Specifications for Power Fuses and Fuse Disconnecting Switches," American National Standards Institute, Inc. New York, 1981.

NRC FORM 335
(2-89)
NRCM 1102,
3201, 3202

U.S. NUCLEAR REGULATORY COMMISSION

BIBLIOGRAPHIC DATA SHEET

(See instructions on the reverse)

1. REPORT NUMBER (Assigned by NRC, Add Vol., Supp., Rev., and Addendum Numbers, if any.)
NUREG-1760

2. TITLE AND SUBTITLE

Aging Assessment of Safety-Related Fuses Used In Low-and Medium-Voltage Applica tions In Nuclear Power Plants

3. DATE REPORT PUBLISHED

MONTH	YEAR
May	2002

4. FIN OR GRANT NUMBER

J2831

5. AUTHOR(S)

R. Lofaro
M. Villaran

6. TYPE OF REPORT

Technical

7. PERIOD COVERED *(Inclusive Dates)*

8. PERFORMING ORGANIZATION - NAME AND ADDRESS *(If NRC, provide Division, Office or Region, U.S. Nuclear Regulatory Commission, and mailing address; if contractor, provide name and mailing address.)*

Brookhaven National Laboratory

Energy Sciences and Technology Department

P.O. Box 5000, Upton, New York 11973-5000

9. SPONSORING ORGANIZATION - NAME AND ADDRESS *(If NRC, type "Same as above"; if contractor, provide NRC Division, Office or Region, U.S. Nuclear Regulatory Commission, and mailing address.)*

Division of Engineering

Office of Nuclear Reactor Regulation

U.S. Nuclear Regulatory Commission

Washington, DC 20555-0001

10. SUPPLEMENTARY NOTES

D. Nguyen, NRC Project Manager

11. ABSTRACT (200 words or less)

An aging assessment of safety-related fuses used in commercial nuclear power pl ants has been performed to determine if aging degradation is a concern for these components in older nuclear power plan ts. This study is based on the review and analysis of past operating experience, as reported in the Licensee Event Report database, the Nuclear Plant Reliability Data System, and the Equipment Performance and Information Exchange System database. In addition, documents prepared by the Nuclear Regulatory Commission that identify significant issues or concerns related to fuses have been reviewed, and one fuse manufacturer was visited to obtain their insights. Based on the results o f the aforementioned reviews, predominant aging characteristics are identified and potential condition monitoring techniques ar e discussed.

12. KEY WORDS/DESCRIPTORS *(List words or phrases that will assist researchers in locating the report.)*

Safety-Related Fuses
Aging Degradation
Potential Aging Mechanism
Potential Age-Related Failures
Low-and Medium-Voltage Applications
Operating Experience
Contaminated Contacts
Loose, Corroded Contacts
Electrical Arcing

13. AVAILABILITY STATEMENT

unlimited

14. SECURITY CLASSIFICATION

(This Page)

unclassified

(This Report)

unclassified

15. NUMBER OF PAGES

16. PRICE

Federal Recycling Program

NUREG-1760

MAY 2002

AGING ASSESSMENT OF SAFETY-RELATED FUSES USED IN LOW- AND
MEDIUM-VOLTAGE APPLICATIONS IN NUCLEAR POWER PLANTS

UNITED STATES
NUCLEAR REGULATORY COMMISSION
WASHINGTON, DC 20555-0001

OFFICIAL BUSINESS
PENALTY FOR PRIVATE USE, $300